CONSEQUENTIAL DAMAGES
OF NUCLEAR WAR

CONSEQUENTIAL DAMAGES OF NUCLEAR WAR

The Rongelap Report

Barbara Rose Johnston and Holly M. Barker

Walnut Creek, CA

LEFT COAST PRESS, INC.
1630 North Main Street, #400
Walnut Creek, California 94596
http://www.LCoastPress.com

Hardback ISBN 978-1-59874-345-6
Paperback ISBN 978-1-59874-346-3

Library of Congress Cataloging-in-Publication Data

Johnston, Barbara Rose.
   Consequential damages of nuclear war : the Rongelap report / Barbara Rose Johnston and Holly M. Barker.
       p. cm.
   Includes index.
   ISBN 978-1-59874-345-6 (hardcover : alk. paper)
   ISBN 978-1-59874-346-3 (pbk. : alk. paper)
   1. Rongelap Atoll (Marshall Islands)—Claims vs. United States. 2. Nuclear weapons—Testing—Health aspects—Marshall Islands—Rongelap Atoll. 3. Nuclear weapons—Testing—Environmental aspects—Marshall Islands—Rongelap Atoll. 4. Radiation victims—Legal status, laws, etc.—Marshall Islands. 5. Radioactive pollution—Marshall Islands. I. Barker, Holly M. II. Title.
KZ238.M37J64 2008
346.03'23—dc22

2008019848

Printed in the United States of America

⊚™ The paper used in this publication meets the minimum requirements of American National Standard for Information Sciences—Permanence of Paper for Printed Library Materials, ANSI/NISO Z39.48—1992.

Cover design by Andrew Brozyna

Cover photo: John Anjain in April 1999, remembering the Bravo shot while standing on the site of his boyhood home. This visit was Anjain's first return to Rongelap since evacuation in 1985. Photo courtesy W. Nicholas Captain.

This book is dedicated
with respect and admiration
to those who can no longer tell their story in person
but whose experiences are partly recounted here:

Alab and former mayor of Rongelap John Anjain
Mr. George Anjain
Ms. Almira Matayoshi

Contents

Illustrations

Prologue

Consequential Damages of Nuclear War

Statement of John Anjain:

Early in the morning of March 1, 1954, sometime around five or six o'clock, American planes dropped a hydrogen bomb on Bikini Atoll. Shortly before this happened, I had awakened and stepped out of my house. Once outside, I looked around and saw Billiet Edmond making coffee near his house. I walked up and stood next to him. The two of us talked about going fishing later in the morning. After only a few minutes had passed we saw a light to the west of Rongelap Atoll. When this light reached Rongelap we saw many beautiful colors. I expect the reason people didn't go inside their houses right away was because the yellow, green, pink, red, and blue colors which they saw were such a beautiful sight before their eyes.

The second thing that happened involved the gust of wind that came from the explosion. The wind was so hot and strong that some people who were outside staggered, including Billiet and I. Even some windows fell as a result of the wind.

The third thing that happened concerned the smoke-cloud which we saw from the bomb blast. The smoke rose quickly to the clouds and as it reached them we heard a sound louder than thunder. When people heard this deafening clap some of the women and children fled to the woods. Once the sound of the explosion had died out everyone began cooking, some made donuts and others cooked rice.

Later some men went fishing, including myself. Around nine or ten-o'clock I took my throw net and left to go fishing near Jabwon. As I walked along the beach I looked at the sky and saw it was white like smoke; nevertheless I kept on going. When I reached Jabwon, or even a little before, I began to feel a fine powder falling all over my body and into my eyes. I felt it but I didn't know what it was.

I went ahead with my fishing and caught enough fish with my throw-net to fill a bag. Then I went to the woods to pick some coconuts. I came back to the beach and sat on a rock to drink the coconuts and eat some raw fish. As I was sitting and eating, the powder began to fall harder. I looked out and saw that the coconuts had changed color. By now all the trees were white as well as my entire body. I gazed up at the sky but couldn't see the clouds because it was so misty. I didn't believe this was dangerous. I only knew that powder was falling. I was somewhat afraid nevertheless.

When I returned to Rongelap village I saw people cooking food outside their cook-houses. They didn't know the powder was very dangerous. The powder fell all day and night long over the entire atoll of Rongelap. During the night people were sick. They were nauseous, they had stomach, head, ear, leg and shoulder aches. People did not sleep that night because they were sick.

The next day, March 2, 1954, people got up in the morning and went down to get water. It had turned a yellowish color. "Oh, Oh" they cried out and said "the powder that fell down yesterday and last night is a harmful thing." They were sick and so Jabwe, the health-aide, walked around in the morning and warned the people not to drink the water. He told them that if they were thirsty to drink coconuts only.

. . . At three o'clock in the afternoon of March 2, 1954 a seaplane from Enewetak Atoll landed in the lagoon of Rongelap and two men came ashore. Billiet and I asked them why they had come to Rongelap and they responded by saying they had come to inspect the damage caused by the bomb. They said they would spend twenty minutes looking at all the wells, cement water catchments, houses and other things. The two men returned quickly to their plane and left without telling anyone that the food, water, and other things were harmful to human beings.

Everyone was quite surprised at the speed with which the men surveyed everything in the island and then returned to their plane. People said maybe we've been really harmed because the men were in such a hurry to leave. Although they said they would look around for about twenty minutes, they probably didn't stay here for more than ten minutes. So in less than ten minutes after their arrival on Rongelap, the two men had already taken off.

. . . On that day we looked at the water catchments, tubs and other places where there was a great deal of water stored. The water had turned a strong yellow and those who drank it said it tasted bitter.

On March 3, early in the morning, a ship and a seaplane with four propellers appeared on Rongelap. Out of the plane came Mr. Oscar [DeBrum] and Mr. Wiles, the governor of Kwajelein Atoll. As their boat reached the shore, Mr. Oscar cried out to the people to get on board and forget about their personal belongings for whoever thought of staying behind would die. Such were the words by which he spoke to them. Therefore, none of the people went back to their houses, but immediately got on the boats and sailed to board the ship that would take them away. Those who were sick and old were evacuated by plane.

. . . At ten o'clock in the morning we left Rongelap for Ailinginae Atoll and arrived there at three in the afternoon. We picked up nineteen people on this atoll and by five o'clock we were on our way to Kwajalein.

On March 4, we arrived on Kwajalein and met the Admiral who then sent us to where we were to stay. A day later, Dr. Conard and his medical team arrived. The doctors were very thorough in checking and caring for our injuries and showed much concern in examining us. The Admiral was also very concerned about our situation and took us in as if we were his own children. His name was Admiral Clark.

Ever since 1954 Dr. Conard has continued to examine the fallout victims on a yearly basis. These visits are very important for all the people on Rongelap and others in the Marshall Islands. These medical examinations are also of great importance for men throughout the world.

. . . From 1959 to 1963 and 1964, after the Rongelapese had returned to Rongelap from Majuro, many women gave birth prematurely to babies which looked somewhat like animals. Women also had miscarriages. During these years many other strange things happened with regard to food, especially to fish in which the fertilized eggs and liver turned a blackish color. In all my forty years I had never seen this happen in fish either on Rongelap or in any of the other places I've been in the Marshall Islands. Also, when people ate fish or [arrowroot] starch produced on Rongelap, they developed a rash in their mouths. This too I had never seen before.

. . . I, John, Anjain, was magistrate of Rongelap when all this occurred and I now write this to explain what happened to the Rongelap people at that time.[1]

[In 1954] the people of Rongelap stayed on Kwajalein for three months and the DOE [Atomic Energy Commission] people removed the Rongelap people to Majuro. The people lived in Majuro for three years and in 1956 the DOE, Trust Territory government and the UN came to Majuro and I went with them to attend a meeting with them at the MII school in Rita. And they told me that it is time that we go back home. And I asked "are we really going home while Rongelap is contaminated?" And the answer that they give me is that "it is true that Rongelap is contaminated but it is not dangerous. And if you don't believe us, well then stay here and take care of yourself."

. . . In 1957 the people returned to Rongelap and the DOE promised that there wouldn't be any problems to the Rongelap people. However in 1958 and 1959 most of the women gave birth to something that was not resembling human beings. There was a woman giving birth to a grape. Another woman gave birth to something that resembles a

---

[1] Excerpts from John Anjain, "The Fallout on Rongelap Atoll and Its Social, Medical and Environmental Effects," ed. and trans. Richard A. Sundt (unpublished manuscript, 1973), on file at the Nuclear Claims Tribunal, Majuro, RMI. Between late 1968 and March 1969, John Anjain wrote a series of articles on the people of Rongelap and their experiences with fallout, evacuation, medical monitoring, and life on a contaminated atoll. His original intent was to publish the series in a Marshallese–English newspaper, a project associated with an adult English-language class being taught on Rongelap by Peace Corps volunteers John and Jean Ranahan. Only a single issue of the paper was produced, and Anjain's initial account was not in that issue. Richard Sundt, also a Peace Corps worker on Rongelap, encouraged Anjain to continue writing his memoir. In 1973 Sundt translated Anjain's writing, developed commentary, and submitted this work in a graduate course on the anthropology of Oceania at the University of Wisconsin–Madison. As Sundt noted in an e-mail to Barbara Rose Johnston on February 29, 2008, "When John wrote his account in 1968 and for long thereafter (so far as I can tell) this was the only written account by a Marshallese voice." An example of Anjain's Marshallese text is included in the appendix.

monkey. And so on. There was a child born at that time and there was no shell covering the top of that child's head.[2]

The American doctors came every year to examine us. Every year they came, and they told us that we were not sick, and then they would return the next year. But they did find something wrong. They found one boy did not grow as fast as boys his age. They gave him medicine. Then they began finding the thyroid sickness.

My son Lekoj was thirteen when they found his thyroid was sick. They took him away to a hospital in America. They cut out his thyroid. They gave him some medicine and told him to take it every day for the rest of his life. The same thing happened to other people. The doctors kept returning and examining us. Several years ago, they took me to a hospital in America, and they cut out my thyroid. They gave me medicine and told me to take it every day for the rest of my life.

A few years after the bomb, Senator Amata Kabua tried to get some compensation for the people of Rongelap. He got a lawyer, and the lawyer made a case in court. The court turned our case down. The court said it could not consider our case because we were not part of the United States. Dwight Heine went to the United Nations to tell them about us. People from the United Nations came to see us, and we told them how we felt. Finally, in 1964, the U.S. Congress passed a bill. The bill gave us money as a payment for our experience. Some of the people spent all their money; some of them still have money in the bank. After we got the money, they began finding the thyroid sickness.

In 1972, they took Lekoj away again. They said they wanted to examine him. They took him to America to a big hospital near Washington. Later, they took me to this hospital near Washington because they said he was very sick. My son Lekoj died after [I] arrived. He never saw his island again. He returned home in a box. He is buried on our island. The doctors say he had a sickness called leukemia. They are quite sure it was from the bomb.

But I am positive.

I saw the ash fall on him. I know it was the bomb. I watched him die.[3]

Statement of Almira Matayoshi:

I was pregnant when they dropped the bomb [Bravo]. I was flown off of Rongelap with the other pregnant women and elderly people. The rest of the people left on the boat. I gave birth to Robert on Ejit, and he was normal. The child I had after Robert, when we had returned to Rongelap, I gave birth to something that was like grapes. I felt like I was

---

[2] "Statement of John Anjain," October 28, 2001, read by Senator Abacca Maddison-Anjain to the Nuclear Claims Tribunal, October 31, 2001.

[3] John Anjain, testimony to the Senate Committee on Energy and Natural Resources, 95th Cong., June 16, 1977.

going to die from the loss of blood. My vision was gone, and I was fading in and out of consciousness. They emergency evacuated me to Kwajalein, and I was sure I was going to die. After the grapes, I had a third child. It wasn't like a child at all. It had no bones and was all skin. When I gave birth they said, "Ak ta men en?" [What is that thing?]. Mama said uror *[a term denoting exacerbation]. It was the first strange child that people had seen. I was the first. That time was the worst in my life. I feel both angry and embarrassed.*[4]

John Anjain is the man whose image appears on the cover of this book. This picture was taken in 1999, as John was describing to the scientists, lawyers, and other experts, assembled with him on his first visit to Rongelap since 1985, what it was that he saw on March 1, 1954, when the hydrogen bomb was detonated, and what he had experienced since. He has his hands up in the air, sketching out the blast as it appeared in the sky, suggesting with this gesture the immenseness of it all.

Almira Matayoshi was a young mother-to-be when caught in the fallout from the 1954 hydrogen bomb. Her experiences in years to come were sadly not unique. Many Marshallese women gave birth to children with indescribable features and characteristics; no Marshallese words existed to describe these children.

What words can possibly communicate what it is like to see and survive such sights? To become increasingly fearful that the intense beauty of your world—the water, the sand, the plants, the soil, the sea, and all the creatures within—has been fundamentally transformed by invisible, untouchable, all-encompassing poison? After years and years of living in a radioactive laboratory as the subject of scrutiny and study, what does it mean to find your fears confirmed—that your favorite foods are taboo, that your loved ones grow old before their time and your children fail to thrive? What does it mean to "survive" downwind from the U.S. proving grounds—where nuclear war was practiced and perfected by Cold Warriors? This book is an effort to answer some of these questions, to tease out the many and varied consequences of the U.S. atmospheric weapons testing program in the Pacific, and to do so in ways that amplify the voice of Marshallese experience.

The Rongelap Report is an expert witness report that was submitted to the Republic of the Marshall Islands Nuclear Claims Tribunal (NCT) in September 2001. It was prepared by Barbara Rose Johnston and Holly M. Barker at the request of NCT Public Advocate Bill Graham, with funding provided by the NCT. The report served as evidence in the Nuclear Claims Tribunal hearing on hardship, pain, suffering, and consequential damages experienced by the people of Rongelap,

---

[4] Almira Matayoshi, interview by Holly M. Barker, Honolulu, June 13, 2001.

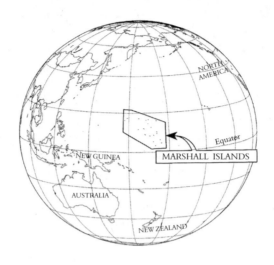

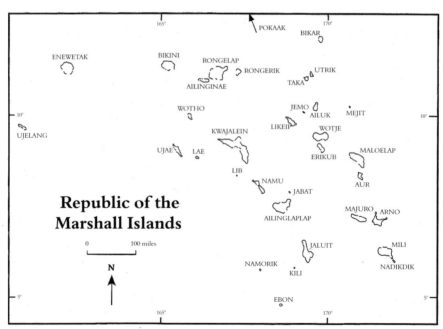

Map 1. The Marshall Islands

Rongerik, and Ailinginae as a result of the actions and activities of the U.S. nuclear weapons testing program. The hearing took place in the fall of 2001 in Majuro, the capital city of the Republic of the Marshall Islands, with the involvement and testimony of the Rongelap people, including Almira Matayoshi and John Anjain, whose poignant memories reprinted above stand in sharp contrast to the public version of events reported in the media. The report is reproduced here with minor editing done for readability and context.

The Rongelap Report tells the story of the myriad of changes that occur to a community whose lives and lands are heavily contaminated with radioactive fallout. In 1946, after evacuating the people of Bikini and nearby atoll communities in the Marshall Islands, the United States detonated two atomic weapons: the same type of bomb that was dropped on Nagasaki in 1945. In 1947 the United Nations designated the Marshall Islands a U.S. trust territory. Over the next eleven years, this U.S. territory played host to another sixty-five atmospheric atomic and thermonuclear tests. The largest of these tests, code named Bravo, was detonated on March 1, 1954. This 15-megaton hydrogen bomb was purposefully exploded close to the ground. It melted huge quantities of coral atoll, sucking it up and mixing it with radiation released by the weapon before depositing it on the islands and inhabitants in the form of ash, or radioactive fallout. The wind was blowing that morning in the direction of inhabited atolls, including Rongelap and Utrik, some 100 and 300 miles from the test site at Bikini. The Marshallese communities on Rongelap, Ailinginae, and Utrik atolls, U.S. servicemen on Rongerik Atoll (weathermen who were monitoring winds and fallout),[5] and the twenty-three-man crew of the Japanese fishing vessel *Fukuryu Maru* (Lucky Dragon) received near-lethal doses of radiation from the Bravo event.[6]

---

[5] Twenty-eight weathermen on Rongerik were exposed to Bravo fallout, and their conditions were monitored as part of Project 4.1. After three months, they were released for duty. Attention to the question of the long-term health effects of fallout on these servicemen did not occur until the threat of lawsuits from the Marshallese and from atomic veterans (some 210,000 servicemen whose duty during atmospheric weapons tests resulted in exposure to radiation). Public Law 97–72, passed in 1981, authorized free health care for veterans whose health problems are a result of exposure during war to ionizing radiation or Agent Orange. In 1982 four air force veterans exposed during Operation Castle tests filed lawsuits challenging the government's explanation of how people came to be exposed. In 1983 the U.S. government announced that Public Law 97–72 also covers atomic veterans whose exposure resulted from service during atmospheric weapons tests (1945–1962). Access to health care requires proof of exposure. In 1984 the navy distributed health questionnaires to atomic veterans and, some thirty years after initial exposure, reassessed its records and issued estimated doses to Rongerik service personnel. See J. Goetz et al., *Analysis of Radiation Exposure—Service Personnel on Rongerik Atoll, Operation Castle—Shot Bravo* (McClean, VA: Science Applications International Corp., 1987), http://www.dtra.mil/documents/rd/DNATR86120.pdf (accessed February 10, 2008).
[6] This Japanese exposure resulted in an international outcry that further fueled the growth of a worldwide antinuclear movement. For a Web-based overview of the *Lucky Dragon* experience, see

International protests and calls for a ban on nuclear weapons testing prompted the U.S. government to publicly acknowledge the incident and accept liability. The Marshallese filed an April 20, 1954, complaint to the United Nations Trusteeship Council:

We, the Marshallese people feel that we must follow the dictates of our consciences to bring forth this urgent plea to the United Nations, which has pledged itself to safeguard the life, liberty and the general well being of the people of the Trust Territory, of which the Marshallese people are a part.

. . . The Marshallese people are not only fearful of the danger to their persons from these deadly weapons in case of another miscalculation, but they are also very concerned for the increasing number of people who are being removed from their land.

. . . Land means a great deal to the Marshallese. It means more than just a place where you can plant your food crops and build your houses; or a place where you can bury your dead. It is the very life of the people. Take away their land and their spirits go also.[7]

In response to this petition the United States assured the General Assembly of the United Nations:

The fact that anyone was injured by recent nuclear tests in the Pacific has caused the American people genuine and deep regret. . . . The United States Government considers the resulting petition of the Marshall Islanders to be both reasonable and helpful. . . . The Trusteeship Agreement of 1947 which covers the Marshall Islands was predicated upon the fact that the United Nations clearly approved these islands as a strategic area in which atomic tests had already been held. Hence, from the onset, it was clear that the right to close areas for security reasons anticipated closing them for atomic tests, and the United Nations was so notified; such tests were conducted in 1948, 1951, 1952 as well as in 1954. . . . The question is whether the United States authorities in charge have exercised due precaution in looking after the safety and welfare of the Islanders involved. That is the essence of their petition and it is entirely justified. In reply, it can be categorically stated

---

"Third Radiation Exposure—The Lucky Dragon No. 5 and Hiroshima," *Hiroshima Peace Museum*, 2005, http://www.pcf.city.hiroshima.jp/frame/Virtual_e/exhibit_e/exhi04_2.html (accessed February 11, 2008).
[7] Petition from the Marshallese People Concerning the Pacific Islands: Complaint Regarding Explosions of Lethal Weapons within Our Home Islands, to United Nations Trusteeship Council, April 20, 1954. In its response to this petition the United Nations Trusteeship Council stated, "The Administering Authority adds that any Marshallese citizens who are removed as a result of test activities will be reestablished in their original habitat in such a way that no financial loss would be involved." United Nations Trusteeship Council, Petitions Concerning the Trust Territory of the Pacific Islands, July 14, 1954, 5.

that no stone will be left unturned to safeguard the present and future well-being of the Islanders.[8]

The United States promised the Marshallese and the United Nations General Assembly that "Guarantees are given the Marshallese for fair and just compensation for losses of all sorts."[9]

These guarantees worked: the United States was able to continue its atmospheric weapons testing program in the Marshall Islands through 1958 and at the Nevada Test Site through 1963, when the United States, Great Britain, and the Soviet Union finally signed on to a limited test ban treaty.

The United States has not, however, fully lived up to its promises to the United Nations or the Marshallese people to safeguard their well-being. As documented in the previously classified studies cited in our expert witness report, atmospheric weapons testing in the Pacific resulted in considerable human and environmental harm.

Global atmospheric nuclear weapons tests released numerous radioisotopes and dangerous heavy metals. An estimated 2 percent of the radioactive fallout was iodine-131, a highly radioactive isotope with an 8-day half-life. The nuclear war games conducted by the United States in the Marshall Islands released some *8 billion* curies of iodine-131.[10] To place this figure in broader context, over the entire history of nuclear weapons testing at the Nevada Proving Grounds, some *150 million* curies of iodine-131 were released, and varying analyses of the Chernobyl nuclear power plant disaster estimate an iodine-131 release of *40 to 54 million* curies.[11] Much of the iodine-131 released in the Marshall Islands was the by-product of the March 1, 1954, Bravo test detonation of the hydrogen bomb. Designed to produce

---

[8] Mason Sears, U.S. representative to the Trusteeship Council, statement to the United States Mission to the United Nations. Press release 1932, July 7, 1954, http://worf.eh.doe.gov/data/ihp1d/400107e.pdf (accessed February 5, 2008).

[9] Frank E. Midkiff, high commissioner of the Trust Territory of the Pacific Islands, statement to the United States Mission to the United Nations. Press release 1932, July 7, 1954, http://worf.eh.doe.gov/data/ihp1d/400107e.pdf (accessed February 5, 2008).

[10] In 1998 the Centers for Disease Control (CDC) estimated some 6.3 billion curies of iodine-131 had been released in the Marshall Islands a result of nuclear weapons tests. The record of radioactive release from atmospheric weapons tests in the Marshall Islands was later reassessed, with the finding that the CDC value "appears too low by at least 32% and possibly by as much as 42%" (Steve Simon, personal communication with RMI Nuclear Claims Tribunal chairman Oscar DeBrum, August 23, 1999, cited in Judge James H. Plasman, testimony to the House Committee on Resources and the Committee on International Relations Subcommittee on Asia and the Pacific, July 25, 2007, http://www.yokwe.net/ydownloads/052505plasman.pdf (accessed February 5, 2008).

[11] See Steven L. Simon, Andre Bouville, and Charles E. Land, "Fallout from Nuclear Weapons Tests and Cancer Risks: Exposures 50 Years Ago Still Have Health Implications Today That Will Continue into the Future," *American Scientist* 94 (January–February 2006): 48–57.

and contain as much radioactive fallout in the immediate area as possible, in order to create laboratory-like conditions, Bravo unleashed as much explosive yield as one thousand Hiroshima-sized bombs. Communities living downwind from the blast, especially the Rongelap community, were acutely exposed to its fallout.

Evacuated three days after the blast, the people of Rongelap spent three months under intense medical scrutiny as human subjects in Project 4.1. They spent three years as refugees and were returned to their still-contaminated atoll in 1957 with assurances that their islands were now safe. They lived on Rongelap for another twenty-eight years and as the closest populated atoll to the Pacific Proving Grounds, they were exposed to additional fallout from another series of nuclear tests in 1958. While living on Rongelap, the community was visited annually, and later biannually, by U.S. government scientists and medical doctors conducting follow-up studies begun under Project 4.1. Researchers collected fish, plants, soil, and human body samples to document the presence of radioisotopes deposited from sixty-seven tests, the movement of these isotopes through the food chain and the human body, and the adverse health impact of this radiation on the human body. The community left Rongelap in 1985 after receiving information from some U.S. scientists that confirmed their long-held fears that their ancestral homeland was contaminated with radiation at levels that posed a serious risk to their health. Today, the Rongelap community lives in exile, largely on borrowed or rented lands in Kwajalein and Majuro atolls. Recent efforts to remediate fallout hazards on areas of some islands and to rebuild homes and community structure on the island of Rongelap suggest that the community may, someday soon, have the choice of returning home. Whether or not remediation is successful and people decide to return remains to be seen.

The Rongelap Report examines the nuclear weapons testing program and its effect on host communities from the point of view of the people of Rongelap. Their recent history is sharply defined by the disastrous events of fallout, acute and chronic exposure to radiation, evacuation in 1954, exile, resettlement in 1957, evacuation in 1985 and again for the past two decades, and struggles to address the many problems of life in exile. The people of Rongelap are not the only nuclear nomads created by the actions of military and nuclear powers over the past six decades. They are, however, one of the most studied communities, and there is much that the world can learn from their experience.

Following their acute exposure in 1954, the people of Rongelap, with residents from a number of other communities, were enrolled in a medical research program sponsored by the Atomic Energy Commission. The program was designed to document the movement of radiation through the atmosphere, food chain, and human body, with the goal of understanding the long-term effects of human exposure to ionizing radiation. This biomedical research was conducted by Brookhaven National Laboratory with monies appropriated by the U.S. Congress

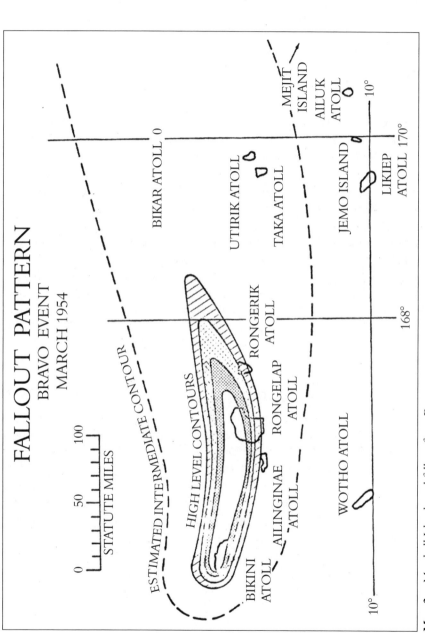

Map 2. Marshall Islands and fallout from Bravo

for the health of the Rongelap people. However, rather than investments in local health infrastructure, funds were used to periodically transport medical staff and supplies from the United States to the Marshall Islands for brief examinations of the "exposed" and "control" populations; to analyze the samples that were collected; to occasionally treat conditions that were defined as radiogenic in nature; and, in later years, to acquire and supply a ship with the necessary technology to conduct whole-body counting, x-ray, and other laboratory procedures. Some of the residents who developed thyroid tumors and other radiogenic conditions were brought to the United States for study and surgical removal of the thyroid gland. When the U.S. government states that it has provided millions of dollars to the Marshall Islands for issues related to the weapons testing, it does not mention that enormous portions of this money went into advancing U.S. scientific interests, not into services for the people.

Over the years, U.S. scientists added to the research program "control" subjects, including people on Rongelap who were not present during the Bravo test, people on the nearby atoll of Utrik, people on Likiep (another populated atoll in the northern Marshall Islands), and people on Majuro. Control subjects were typically selected to match the acutely exposed by age and sex, and scientists studied these people in many instances for four decades. Comparative studies documented increases in thyroid disorders, stunted growth in children, and increases in many forms of cancer and leukemia, cataracts, and other radiation-related illnesses.

For four decades, U.S. government scientists returned to the Marshall Islands to conduct exams and collect blood, tissue, bone marrow, teeth, and other samples. These studies generated a broad array of scientific findings, including the recognition that not only can acute exposures to radiation stimulate short-term effects but that late effects can emerge years and decades following the initial exposure. For example, by studying the Marshallese population, scientists found that radioiodine (I-131) adheres to and accumulates in the thyroid, stimulating the production of benign and cancerous nodules and interfering with the production of hormones, leaving pregnant women and children especially vulnerable. They also discovered that people who were not exposed to an acute level of ionizing radiation but were exposed to low-levels on a daily basis because they lived in an area contaminated by fallout also developed thyroid and other radiogenic problems. The lessons learned by scientists included an awareness of the many complicated ways that radiation adversely affects the human body.[12] In today's world—where

---

[12] In the 1980 review of medical findings from twenty-six years of AEC-sponsored research, some 260 reports and publications resulting from the Marshallese studies are cited. See R. A. Conard, *Review of Medical Findings in a Marshallese Population Twenty-Six Years After Accidental Exposure to Radioactive Fallout* (Upton, NY: Associated Universities, 1980). More has been published in the years since. See, for example, E. T. Lessard, R. P. Miltenburger, S. H. Cohn, S. V. Musolino, and

uranium mining occurs at historic levels, where depleted uranium is widely used in military training and war, and where nuclear power and weapons production are again on the agendas of the world's nations—these lessons have currency. The experiences of the people of Rongelap, whose lives were transformed not only by acute exposure but also by chronic exposure to low-level radiation, should be read as a timely, cautionary tale.

The Rongelap Report details how much of the scientific study in the Marshall Islands, especially the research conducted over the first three decades of the program, occurred with top-secret classification, and how this biomedical research was often conducted without meaningful informed consent. The classified nature of this research had profound effects within the Marshal Islands and within the broader scientific research community. Research protocols, data, and findings were restricted to those with security clearance. Patients, and later the Marshall Islands government, were denied access to medical records generated by this research. The research agenda itself was shaped to meet U.S. military and scientific research objectives rather than the personal health needs of the affected population. The pressing question for the U.S. government was how to document and interpret the Marshallese experience in ways that might predict the consequences for U.S. troops or U.S. citizens exposed to radiation in the event of nuclear war. Marshallese health concerns, especially worries that radiation from fallout remained in their environment, poisoning their food and their bodies, were often ignored.

The culture of secrecy that characterized biomedical research in the Marshall Islands facilitated efforts to shape public opinion on the safety of the nuclear weapons testing program. Scientific findings were cherry-picked: those studies released to the public were carefully selected; conclusions were carefully worded to support the contention that exposed communities suffered no lasting effects from their exposure and that their exposure presented no threat to the health of subsequent generations. As mentioned above, such actions were taken to counter calls within the United Nations to establish a ban on nuclear weapons testing; to calm local and regional complaints that exposure to radiation was producing a wide array of untreated health effects, especially reproductive effects; and to

---

R. A. Conrad, "Protracted Exposure to Fallout: The Rongelap and Utirik Experience," *Health Phys* 46 (1984): 511–27, and, T. Takahashi, M. J. Schoemaker K. R. Trott, S. L. Simon, K. Fujimori, N. Nakashima, A. Fukao, and H. Saito, "The Relationship of Thyroid Cancer with Radiation Exposure from Nuclear Weapon Testing in the Marshall Islands," *Journal of Epidemiology* 13, no. 2 (March 2003): 99–107.

reduce the economic liability of the U.S. government in meeting its obligations to its former territory.[13]

The Rongelap Report differs from other efforts that document for a public audience the nuclear weapons testing history and related experiences of the Marshallese people. Other published accounts typically rely upon government-controlled interpretations or upon survivor memories and understandings. Because so much of the data on nuclear fallout and the movement of radionuclides through the food chain and the human body were classified, and because the funding and priorities of human environmental research was controlled by the United States, it has been extremely difficult to produce evidence that corroborates personal testimony. Thus, over the years, Marshallese complaints have been easily dismissed as anecdotal accounts that fly in the face of scientific findings.

On a number of occasions since 1954, Rongelap residents have sent formal letters protesting conditions and health problems to the Atomic Energy Commission (AEC), Marshallese politicians, and members of the Congress of Micronesia. Some of these complaints involved health problems occurring outside of the recognized "exposed" atoll populations of Bikini, Enewetak, Rongelap and Utrik. Many complaints involved the sudden experience of previously unknown illnesses and conditions, especially reproductive health problems such as infertility and the birthing of grossly deformed babies. These complaints are presented again in *The Rongelap Report,* as are the carefully constructed replies that were sent back, assuring people that the problems they were experiencing were normal, were to be expected in a small island population, and had nothing to do with the nuclear weapons testing program. Close examination of the record of debate within the AEC after receipt of these complaints reveals a public and a private recounting of events in the RMI. In sharp contrast to the carefully constructed, placating assurances the AEC sent to the Marshallese people, the AEC noted in its internal communications extensive radioactive contamination of the terrestrial and marine food chain, and recontamination of Rongelap following tests in 1958. Reports and transcripts predicting the human effects of radiation exposure in the Marshall Islands include

---

[13] See, for example, Dr. Robert Conard, Brookhaven National Laboratory, letter to Dr. James L. Liverman, Division of Biological and Medical Research, Energy Research and Development Administration, March 1, 1977, with suggested statements to use in response to letters from the people of Rongelap and Utrik, http://worf.eh.doe.gov/data/ihp1a/1584_.pdf (accessed February 10, 2008). For more detailed discussion and case-specific essays on U.S. and Soviet Cold War militarism and secrecy, and the related caustic impacts on public health and the construction of science, see Barbara Rose Johnston, ed., *Half-lives and Half-truths: Confronting the Radioactive Legacies of the Cold War* (Santa Fe, NM: SAR Press, 2007). For a brief look at the Rongelap experience, see Holly Barker, *Bravo for the Marshallese: Regaining Control in a Post-Nuclear, Post-Colonial World* (Belmont, CA: Thomson, Wadsworth Publishers 2004).

candid discussions of adverse impacts on human fertility and reproduction. The annual and biannual medical surveys of the Rongelap population carefully record rates of miscarriage and congenital defects (although no research program was developed to systematically study or treat these concerns).[14] On a few occasions when deformed children were born to women on Rongelap, scientists flew in to examine and photograph them. And public health officers throughout the Trust Territory were asked to pay careful attention and to document reproductive abnormalities. For example, Pacific Trust Territory records include an account by a public health officer of his visit to Wotho in 1957:

Oct. 5, 1957—I went to the shore at 07:30 this morning to perform a biopsy on [a female patient's] back. I also visited the patient who has cancer of the breast. This case is hopeless so I told [the health aide] to treat her symptomatically. . . . There is another case on the island of Wotho. A girl who is about ten (10) months old and has no vagina. I wanted to bring the child to Majuro but the parents were not ready to come on this trip. I told them to be coming on next field trip.[15]

Testimony from Rongelap survivors indicates that, following their evacuation in 1954, they were told to expect unusual births and higher incidences of miscarriage in the months and years following their exposure. Most Marshallese, however, were not given such advice and were absolutely unprepared to understand and deal with the varied and extreme defects in their children, previously unknown conditions that occurred in the months and years following the nuclear weapons tests.

As the decades passed, people experienced a growing incidence of adverse health effects, most notably the late onset of thyroid cancer and stunted growth and retardation in children in "exposed" as well as "control" populations. These health problems fed concerns that Rongelap Atoll was still dangerously contaminated and

---

[14]A reassessment of I-131 exposure and its health effects in the Marshall Islands, including adverse effects of hypothyroidism from I-131 exposure during pregnancy, was conducted by Hans Behling in 2006. Behling summarizes the ample evidence of adverse pregnancy outcome in the Brookhaven National Laboratory records and notes that despite "all the data that had been collected by BNL scientists that included thyroid dose estimates in hundreds to more than 1,000 rads, clinical evidence of thyroid pathologies/ hypothyroidism, and dietary deficiencies in iodine (that was likely further exacerbated by deficiencies in iron, selenium, and vitamin A), the role of impaired thyroid function among pregnant females was never considered by BNL scientists as a risk factor that might explain the observed adverse pregnancy outcomes among the exposed population groups." U. Hans Behling, *An Assessment of Thyroid Dose Models Used for Dose Reconstruction*, vol. 2, *A Critical Assessment of Historical Thyroid Dose Estimates for Marshallese Exposed to Test Bravo Fallout* (Vienna, VA: S. Cohen and Associates, May 2007).
[15]Public Health Department, *WFT Report* (Trust Territory of the Pacific Islands, October 15, 1957); medical field trip officer Isaac K. Lanwi to the TTPI district administrator. TTA microfilm roll 994, Trust Territory of the Pacific Islands Archives, University of Hawaii–Manoa. In this quote, names have been removed to protect the privacy of individuals.

posed a significant hazard to occupants, a fact that became evident in the restudy of radiological conditions in the northern Marshall Islands in 1978. The results of this survey and the input of a few independent foreign experts led the Rongelapese to evacuate their homes in 1985, with the assistance of the Greenpeace ship *Rainbow Warrior* on what proved to be its final voyage in the Pacific.[16] The evacuation of Rongelap occurred without the assistance or approval of the U.S. government. The restudy confirmed that much of the northern Marshall Islands was indeed still contaminated and that some areas would not be habitable without extensive remediation for at least twenty-five thousand years.

Nuclear Claims Tribunal Proceedings

In 1986, after years of negotiations and the threat of some $7.1 billion in damage claims making their way through the U.S. court system, the United States and the Republic of the Marshall Islands signed a Compact of Free Association, releasing the U.S. government from pending legal claims through the establishment of a compensation trust fund. The Compact of Free Association requires the United States to continue efforts to adequately address the full range of damages and injuries resulting from the testing program. Section 177 of the compact outlines responsibilities for monitoring the environment and human health effects of radiation from the nuclear weapons tests in the northern Marshall Islands (Bikini and Enewetak, the two ground-zero locations and Rongelap and Utrik atolls, the two communities enrolled in the Project 4.1 biomedical study). An additional provision of section 177 enables the Republic of the Marshall Islands to petition the U.S. Congress for additional compensatory funds should conditions change or new information come to light. Congress set aside $150 million to fund the provisions of the initial compact, which established a compensation trust fund with funds administered through a Nuclear Claims Tribunal that receives claims and issues awards for personal injury and property damage.

The *Rongelap Report* is one of many expert witness reports prepared over the years in support of NCT deliberations. The NCT is an administrative court governed by Marshallese and U.S. law. It is set up to receive personal-injury and property-damage claims, produce full and final judgments in all claims, and issue payments from the trust fund established by the U.S. Congress. Three judges sit on the Nuclear

---

[16] On July 10, 1985, Greenpeace's ship *Rainbow Warrior* was sunk while at dock in Auckland, New Zealand, by two explosive devices placed on its hull by French commandos working under the authorization of French president François Mitterrand. The detonation killed the photographer Fernando Pereira. The French sabotage was an attempt to disrupt Nuclear Free Pacific protests against French nuclear testing at Mururoa Atoll in the Tuamotu Archipelago of French Polynesia. See David Robie, *Eyes of Fire: The Last Voyage of the Rainbow Warrior,* revised ed. (Auckland: South Pacific Books, 2005).

Claims Tribunal. Claims are developed and presented by the Public Advocate, who with his office staff assists individuals and communities in the preparation of claims. Claims are reviewed by the Defender of the Fund, who may present a case in defense of the fund against personal-injury claims but may also admit claims for clearly compensable medical conditions. The personal-injury program of the RMI Nuclear Claims Tribunal is modeled after the 1990 U.S. congressional program for Downwinders exposed to the Nevada tests. Because it is impossible to determine if a cancer or illness linked to radiation exposure is a direct result of exposure to radionuclides, the tribunal program, like U.S. programs, presumes that if a claimant was either alive or in utero during the testing program, and if he or she contracts an illness highly associated with radiation exposure, then the condition is a result of that exposure. The property-damage program considers damage from nuclear weapons testing, including loss of access or use, cost to restore, and the pain, suffering, and hardships that are the "consequential damages" of these losses.[17]

The *Rongelap Report* is a component of the third major property-damage claim presented to the NCT. Earlier claims presenting the case of property damage on Bikini and Enewetak atolls were developed within a Western property-rights framework. In these cases, property damage as a result of nuclear weapons testing was largely defined in terms of values associated with the loss of use of dry land, with the assumption that the economic value of using dry land (established through the record of rental agreements set up by the U.S. military) is the sum total of value. For many reasons, this approach did not do justice to the particular problems of the Rongelap experience. Personal injuries were compensated by the NCT in the event that radiogenic cancers occurred and were documented, but people did not "lose" their land. Rather, they lost the safe use of their lands. The methods used to develop prior property-damage claims did not allow consideration of the broader injuries resulting from the classified medical research program, the pain and problems of living as a stigmatized member of the "bombed" community, the many varied instances of pain, illness, and suffering resulting from years of living in a heavily contaminated setting, or the hardships and losses of having to again evacuate their homelands when the U.S. government failed to acknowledge to the Rongelapese the lingering and dangerous contamination they were living with.

Thus, in *The Rongelap Report,* we ask the Nuclear Claims Tribunal to first consider the value of what was a preexisting, self-sufficient way of life, before moving to

---

[17] Additional detail on the history of the tribunal and a thorough review of tribunal proceedings and judgments issued through 2002 is contained in an independent audit conducted by former U.S. attorney general Dick Thornburgh. See Dick Thornburgh, Glenn Reichardt, and Jon Stanley, *The Nuclear Claims Tribunal of the Republic of the Marshall Islands: An Independent Examination and Assessment of Its Decision-Making Process* (Washington, DC: Kirkpatrick and Lockhart, 2003), http://www.bikiniatoll.com/ThornburgReport.pdf (accessed February 5, 2008).

the questions of how best to identify and assess damage and repair injury. We ask the NCT to consider a chain of nuclear weapons testing events that resulted in multiple abuses of person, property, and fundamental dignities. And drawing upon Marshallese values and norms, we argue for a redefinition of the compensation principle employed in prior cases—from a model of compensation for damage and a loss of individual property rights to a broader model that considers and addresses the community damages associated with the loss of a way of life.

Classified Science and Government Transparency

The Rongelap Report and its holistic approach to the chain of events, record of injury, and assessment of consequential damage offers a tiny sense of what it must be like for an island population to survive downwind of a nuclear war. This report takes on added relevance when considering that what was known by the general public and Marshallese officials when the Compact of Free Association was negotiated and the Nuclear Claims Tribunal was established. At that time, much of the scientific record was classified. The Marshallese were never fully briefed on the nature of the nuclear weapons testing program and the full extent of its damages. This inequitable access to fundamental information has severely hampered Marshallese efforts to achieve a meaningful and comprehensive remedy. For example, to this day, the United States acknowledges in its compensatory programs the obligation to address nuclear-weapons-related damage to property and people in only four atolls: Bikini, Enewetak, Rongelap and Utrik. The U.S. documentary record tells another story: a 1955 survey, declassified in 1994 and released to the RMI in 1995, reports fallout from the 1954 Bravo test occurring at hazardous levels on twenty-eight atolls throughout the Marshall Islands. The entire nation, not simply the four atolls, is downwind, and the whole country has been adversely affected by nuclear weapons.[18]

---

[18] See Alfred Breslin and Melvin E. Cassidy, *Radioactive Debris from Operation Castle, Islands of the Mid-Pacific* (New York: New York Operations Office, United States Atomic Energy Commission, January 18, 1955), http://www.yokwe.net/ydownloads/RadioactiveDebrisCastle.pdf (accessed February 5, 2008). This document was declassified by the U.S. Department of Energy in 1994 as part of ACHRE review. A copy was provided to the Republic of the Marshall Islands in 1995. The report shows that radiation fallout doses were measured at sites throughout the Marshall Islands atolls following each of the six tests conducted in 1954. The document reports significant fallout on twenty-eight atolls, of which twenty-two were populated during Operation Castle (March 1 through May 14, 1954). No monitoring was reported for the islands of Mejit, Lib, and Jabat. The fact that, regardless of location in the Marshall Islands, all residents were exposed to substantial levels of radioactive fallout as result of the atmospheric weapons tests was confirmed in Hans Behling, John Mauro, and Kathy Behling, *Final Report: Radiation Exposures Associated with the U.S. Nuclear Testing Program for Twenty-one Atolls/Islands in the Republic of the Marshall Islands* (McLean, VA: S. Cohen and Associates, 2002).

Map 3. Comparing Nevada Test Site and Bikini Atoll

While elements of the Rongelap story have appeared in various newspaper accounts, articles, documentaries, and books, the telling of the human dimensions of this story has largely relied upon anecdotal accounts of the key actors: Marshallese survivors, U.S. military veterans, and U.S. scientists.[19] Only recently have we

_____

[19] For a survivor point of view, see the many instances in which Marshallese citizens testified in U.S. Senate and congressional hearings. See, for example, Jeton Anjain, testimony to the House Committee on Interior and Insular Affairs regarding the safety of Rongelap Atoll, 1989. Jeton Anjain

seen the material evidence that substantiates anecdotal accounts come to light via the declassification of materials documenting the military research agenda, protocols, experiments, and findings concerning the nature of radioactive fallout, its movement through the food chain and human body, and its long-term effects. In 1993, in response to series of newspaper articles describing Cold War–era radiation experimentation involving U.S. subjects, President William Jefferson Clinton ordered a review of files and a declassification of documents pertaining to human radiation experiments in all government agencies. As a result, a trickle and then a flood of documents began to appear on the doorstep on the Marshall Islands Embassy in Washington D.C. In 1994 the Advisory Commission on Radiation Experimentation (ACHRE) assessed the available record and concluded that the Marshallese had indeed served as research subjects in several experiments involving radioisotopes but found no conclusive evidence that the long-term research program was an example of human radiation experimentation.[20] By this time, a truckload of declassified documents had been located by U.S. government agencies and dumped on the RMI Embassy doorstep. Holly Barker reviewed some of these as part of her effort to determine what the U.S. government knew but had not told the RMI government and to develop an ethnographic record for the Marshallese government. Following ACHRE's publication of its findings, the historical documents reviewed by ACHRE were scanned and placed on the word-searchable Human Radiation Experiments (HREX) website and were made accessible to the public in November 1996. As documents continued to be located and released, they were added to a database managed by the DOE. By late 1998, when Public Advocate Bill Graham asked us to advise the Nuclear Claims Tribunal on culturally appropriate ways to

---

initially trained as a dentist, assisted the Brookhaven medical survey, and became health minister and then senator in the Marshall Islands. He died of cancer in 1993. It was through his efforts that the people of Rongelap were evacuated with the assistance of Greenpeace's *Rainbow Warrior* in 1985. His testimony and advocacy on behalf of the Marshallese led the Golman Foundation to award him the Goldman Environmental Prize in 1992. See http://www.goldmanprize.org/node/72. For a visual account of the Rongelap story contrasting the lived experience with the formal military framing of that experience, see Dennis O'Rourke's film *Half Life: A Parable for the Nuclear Age* (Los Angeles: Direct Cinema, 1986). Daily life in the midst of nuclear war games is captured in Michael Harris's book *The Atomic Times: My H-bomb Year at the Pacific Proving Ground, a Memoir* (New York: Ballantine Books, 2005). The perspectives and sentiments of a government scientist in charge of the biomedical research program can be found in Robert Conard, *Fallout: The Experiences of a Medical Team in Care of a Marshallese Population Accidentally Exposed to Fallout Radiation* (Upton, NY: Brookhaven National Laboratory, 1992). For a critical assessment of how public perception of the atomic age was shaped and manipulated via U.S. government control of information and a publishing relationship with the *New York Times,* see Beverly Ann Deepe Keever, *News Zero: The New York Times and The Bomb* (Monroe, ME: Common Courage Press, 2004).
[20] Advisory Committee on Human Radiation Experimentation, *The Human Radiation Experiments: Final Report of the President's Advisory Committee* (Washington, DC: U.S. Government Printing Office, 1995), chapter 12: "The Marshallese," http://www.hss.energy.gov/healthsafety/ohre/roadmap/achre/chap12_3.html (accessed February 5, 2008).

value the damages and losses experienced by the people of Rongelap as a result of the weapons testing program, another ten boxes of documents had been declassified and turned over to the Marshall Islands government. By the end of 1999, DOE had turned over some seventy-seven boxes of declassified information on the nuclear weapons testing program that had been previously withheld from the Marshallese. It was this expanded set of declassified records that was used to build the chain of events and support consequential damage findings in *The Rongelap Report*.

The Rongelap Report was prepared under the very best of research conditions. We had complete access to and the collaborative assistance of the affected community. Our methods were developed and findings critically evaluated by a scientific peer-review committee consisting of environmental and medical anthropologists, ecologists, health physicists, physicians, psychologists, resource economists, appraisers, historians, and lawyers. We had the transcripts of oral histories conducted by Holly Barker with hundreds of radiation survivors over a five-year period. And we had access to the historical documents and scientific reports housed at the Republic of the Marshall Islands Embassy in Washington, D.C., and the Nuclear Claims Tribunal in Majuro, RMI.

Perhaps most importantly, we had the use of a word-searchable research engine to easily locate in the many tens of thousands of declassified documents pertaining to the history of nuclear weapons testing and human radiation experimentation those particular memos, letters, transcripts, trip reports, and scientific studies that might address specific questions, issues, or actions. We could find the needle in the haystack. We could demonstrate, for example, that not only were people complaining about poisoning after consuming locally grown food and locally caught fish, but that the Atomic Energy Commission knew of these complaints and that their scientists had taken samples, found high levels of radioiron (Fe-59) and other isotopes, and had then replied to the Marshallese with placating statements blaming ill health effects on food preparation rather than acknowledging the radioactivity in the food itself. We could search for each and every document that included the term *medical matters* or *blood samples* or *bone marrow* or *radioisotope* and in so doing we could patch together a record of invasive medical sampling that spanned decades, with "exposed" and "control" subjects providing information used in a host of studies. This access to previously classified data and search-engine tools allowed us to craft a holistic look at the Rongelap experience and for the first time present the anecdotal record as a substantiated body of evidence.

At the time *The Rongelap Report* was accepted by the Nuclear Claims Tribunal in September 2001, the chain of events and supporting documentary evidence were easily verified by accessing the URLs included with each declassified document citation. Access to the declassified documents that support this historical account is a much more difficult matter today, and it is this fact that drove our efforts to publish *The Rongelap Report*. In January 2003, while conducting follow-up research for the Nuclear Claims Tribunal, online access was denied for specific documents the HREX website cited in *The Rongelap Report*. One such denial read, "Your client does not have permission to get URL/tiffs/doe/d714594/d714594a/14601012.tif from this server." Over the next few months, more and more documents became inaccessible. By October 2003, the HREX site was completely shut down, with a simple statement claiming that the Department of Energy lacked the funds to maintain the site. Several months later, the statement was changed to its current (January 2008) wording:

> *The HREX website is currently closed down for two reasons: (1) After the events of September 11, 2001, the Federal Government undertook a review of all information on its websites to determine the appropriateness of the information on the websites. The database for the HREX website is currently undergoing a review in light of the events of 9/11 to determine whether all of the information in the database is appropriate. (2) The HREX website was hosted by antiquated technology. After the review of the information in the database is complete, it will be moved to the OpenNet website which is a platform composed of current technologies. The timing of when the HREX information will be available on OpenNet is unknown, therefore you may wish to periodically visit the OpenNet website at http://www.osti.gov/opennet. Thank you for your patience.*

At this writing, it is unclear how much of the supporting documentation for *The Rongelap Report* has been removed from public access or reclassified. In July 2005, with support from a John T. and Catherine D. MacArthur research and writing grant, Barbara Rose Johnston traveled to Washington, D.C., and confirmed that hard copies of HREX files were available at the National Security Archive, a nonprofit institute at George Washington University. This archive includes only those documents that were available during the time ACHRE operated and not the many documents located and declassified in the years since. A number of key documents cited in this report are still accessible on the DOE Marshall Islands document collection archive (http://worf.eh.doe.gov/). Researchers can also access select nuclear testing and radiation experimentation documents by searching the DOE's Office of Science and Technology Information (OSTI) archive and requesting copies through its Bechtel subcontractor (although requests for documents in 2006 and 2007 have gone unanswered). The archive, and status of human radiation experiment records declassified by other agencies, such as the Department of Defense and the Central Intelligence Agency, after ACHRE ceased operating, is unknown. A number of key HREX documents are no longer accessible through the DOE websites, search engines, or OSTI archive. While copies of all documents cited in *The Rongelap Report* are retained by the authors, the Republic of the Marshall Islands, and the RMI Nuclear Claims Tribunal, public access has been significantly reduced.

This loss of public access to declassified documents occurs at a time when the U.S. government is classifying at historic levels a broad array of information pertaining to the perceived security of the state. Effective governance requires the ability to anticipate and meet the basic needs of the public, and to do so in ways that generate and sustain the trust of the public. Transparency is a key element in securing and maintaining trust, especially when earlier actions have broken that trust. The experience of researching and writing *The Rongelap Report* and participating in the RMI Nuclear Claims Tribunal taught us a great deal about the importance of truth, of forums where truth can be told and responsible parties or their representatives can listen, acknowledge, and concur that grave injustices have occurred. Access to the declassified data, and access in searchable form, allowed us not only to record the varied accounts of the difficulties of life and death in a radioactive nation but also to give power to those accounts by identifying the many documents in the declassified record that substantiated the human history, experiences, pain, and suffering.

Finally, the publication of this report at this time reflects a growing awareness that our understanding of history is easily manipulated and even fabricated, especially when the public loses access to the primary documents that underlie historical accounts. The declassification of the 1990s opened a window of transparency, allowing public access to those documents that confirmed the worst fears about how a government took evil action to ensure a political good—securing the military and economic status of the nation to fight the Cold War in ways that involved horrific abuses of fundamental human rights. Declassification and public scrutiny of historical injustices also represent an opportunity to come to terms with the past, and in doing so to take honest and significant effort toward making amends. Shutting down the HREX website effectively reduces public access to the primary documents that tell a story at odds with current administration policy.

Seeking Meaningful Remedy

The story of Rongelap is one of systemic injury, and inadequate and at times abusive response on the part of the U.S. government. U.S. government activities in the Marshall Islands resulted in profound consequences for the entire nation, unmet U.S. obligations, and an intergenerational responsibility. At this writing, the U.S. government views its responsibility to its former territorial possession, and those people adversely affected by the nuclear weapons testing program, as a set of limited obligations that have in large part been addressed. *The Rongelap Report,* and the broader body of reports emerging from the deliberations of the Nuclear Claims Tribunal, identifies experiences, conditions, and continued problems of the Marshallese that contrast sharply with the U.S. government's contention that it has responded adequately to its obligations.

The U.S. government dehumanized the Marshallese by turning their experiences with nuclear weapons into numbers and statistics, by classifying and restricting public access to the documentation of the human environmental impacts of nuclear militarism, and by trivializing the extent of human environmental damage for political gain. This report returns the voice of the Rongelapese to the forefront and provides the community with the opportunity to explain the effects of the testing program from its point of view.

This report also offers a model for documenting human rights abuse and related environmental crises in ways that suggest or lead to meaningful remedy. Working with affected peoples to document preexisting ways of life, the chain of events, injuries, and consequential damages from injuries in ways that respect community experience, understandings, and priorities provides opportunities to identify and prioritize remedial needs. Collaborative and participatory approaches give broad ownership to both the process and the product, and such efforts are transformative: victims take proactive roles in setting, and ideally achieving, the remedial agenda. Combining such process with a holistic consequential damage assessment produces both a powerful experience and a powerful outcome. Assessing the consequential damages of abuse in holistic, human, environmental fashion allows a more thorough understanding of what was, what occurred, to what effect, and, most importantly, what the affected population defines as its remedial needs: what is needed to heal, survive, and thrive. The radioactive contamination of land results in much more than the simple loss of land. Indeed, such contamination undermines a cultural way of life and results in the loss of a healthy way of life. Thus meaningful compensation must necessarily address both the loss *and* the remedial means to restore a sustainable way of life.

What is meaningful remedy? For the Rongelap people, the notion of remedy involves much more than simple monetary compensation. Remedy involves, first and foremost, honest reconciliation with the past in ways that allow understanding of the ramifications of this past on the present and the future. In sharing their experience with the world, the people of Rongelap share with you, the reader, a sense of what it means to survive nuclear war and their sincere plea: never again.

Acknowledgements

This work is the end result of five decades of Marshallese commitment to understanding the extent of their injuries, documenting and filing complaints, and securing meaningful help. Evidence of this effort to ensure that the world knows the Rongelap story, that true lessons are learned, and that the indignities and abuses are not repeated appear again and again in the documentary record. These words and efforts inspire, and they will not be dismissed. *Kommol tata.*

Initial research and *Rongelap Report* production was supported by funds provided by the Marshall Islands Nuclear Claims Tribunal, Office of the Public Advocate, and staff release time for Holly Barker from her advisory duties at the RMI Embassy. The resulting expert witness report is a public document accepted by the tribunal and filed in its office in Majuro. It is reprinted here with the knowledge and permission of the tribunal. Sadly, its publication appears after the July 23, 2002, death of the honorable Oscar DeBrum, a man whose life as a public servant in the administration of the Pacific Trust Territory and the new Marshallese government, and later as chairman of the Nuclear Claims Tribunal, was devoted to the struggle to alleviate the suffering caused by the nuclear testing program. Publication of the report was delayed until the NCT rendered a final judgment in the Rongelap claim. As discussed in the epilogue, decisions in the Rongelap claim were announced as part of an NCT judgment issued on April 17, 2007.

Research and writing of the framing chapters occurred with the financial support of the John T. and Catherine D. MacArthur Foundation through a research and writing grant to the Center for Political Ecology (2004–05) entitled "Considering the Consequential Damages of Nuclear War Legacies," a Weatherhead Resident Scholar Fellowship at the School for Advanced Research for Barbara Rose Johnston (2006–07), and a grant by the Christensen Fund to the Center for Political Ecology. Again, the Embassy of the Marshall Islands played a helpful role in providing staff release time for Holly Barker.

Since submitting this report to the NCT in September 2001, we have reformatted the text, corrected errors in the text and notes, and updated the Web citations. Complete citations for supporting references are contained in footnotes. A number of the supporting documents submitted with this report remain accessible on the Web, while others do not. We include a list of the supporting documents submitted to the tribunal at the end of this book, and we reprint those supporting documents that are not Web-accessible.

The Rongelap Report uses Marshallese expressions, military references, health physics data, and terms—a great deal of language that may not be familiar to the reader. For this reason, we have compiled terms and their meanings in a glossary that appears at the end of this book.

This work is illustrated with images and maps submitted as supporting evidence in the NCT proceedings, as well as images located in personal and public archives in the years since. We thank Nick Captain for his permission to reprint the John Anjain cover photo. We thank Glenn Alcalay for permission to use his images of the Rongelap evacuation which he took in 1985 while documenting the Greenpeace-assisted resettlement. We thank Giff Johnson for the use of his photos. And we gratefully acknowledge the assistance provided by Bill Graham, Glenn Alcalay, Giff Johnson, and others who helped locate and provide information about photos taken many decades ago.

As mentioned in *The Rongelap Report,* the Marshallese were intensively photographed on each of the sixty-seven medical survey trips and throughout the initial three-month Project 4.1 documentation of radiogenic health effects from acute exposure to fallout. Over the years, exam notes, sampling results, x-ray film, whole-body counter data, and photo records resulted in very large medical files for each human subject in the AEC-funded research conducted by Brookhaven scientists. Several copies these records existed—at Brookhaven Laboratory and at the Trust Territory public health offices; in the 1970s, a portion of the Rongelap medical record was placed in an embassy safe in Japan. As recounted in *The Rongelap Report* and further detailed in the tribunal hearing, around the time that lawsuits were working their way through U.S. courts and the United States began to negotiate terms for independence, a series of fires occurred, reportedly beginning in locked safes or file cabinets and resulting in the loss of medical records and photographs in the Marshall Islands and Japan. Requests from the Marshall Islands government for complete access to medical records and photographs compiled by the AEC and Brookhaven remain unfulfilled. Copies of the images taken by Brookhaven scientists and Trust Territory public health officers of the grossly deformed children born to the people of Rongelap, Utrik, and other Marshall Island populations have been especially difficult to find.[21]

This report makes liberal use of direct quotes from interviews and key witness testimony. Actual names are used where they are a matter of public record. Interviews were conducted with the understanding that testimony was being recorded, would be transcribed, and would be incorporated into the formal report. Thus attributions for direct quotes appear with the informed consent of the Marshallese. In the epilogue, quotes derived from post-hearing interviews are attributed by initials rather than names, to protect the privacy of individuals. Throughout the report, there is some variation in the spelling of the names of people and places: Marshallese names often change over time, and the spelling of names can vary greatly from one source to another.

Suggestions for documenting the traditional way of life and comments on draft findings were provided by a Marshallese advisory committee including Iroij Mike Kabua, Senator Wilfred Kendall, Councilwoman Lijon Eknilang, Councilman

---

[21] Photojournalists from Japan have compiled a record of images capturing some of the health problems experienced in the Marshallese population, arguably as a result of first-, second-, or third-generation exposure to radiation. See, for example, Hiromitsu Toyosaki, *Good-bye Rongelap!* trans. Masayuki Ikeda and Heather Ikeda (Tokyo: Tsukiji. Shokan, 1986) and the photo-documentary work of Kousei Shimada, including *Bikini: The Testimony of Bomb Victims in the Marshall Islands* (JPU Shuppan, 1977). For a narrative description, see James N. Yamazaki, *Children of the Atomic Bomb: An American Physicians Memoir of Nagasaki, Hiroshima, and Marshall Islands,* with Louis B. Fleming (Durham, NC: Duke University Press, 1995).

George Anjain, and handicraft businesswoman Mary Lanwi. The research questions, strategies, and data-collection efforts were supported by the collaborative involvement of Nuclear Claims Tribunal Public Advocate Bill Graham, associate Public Advocate Tieta Thomas, and Marshallese anthropologist Tina Stege. In an early phase of this work, in 1999, Stuart Kirsch provided background information on land-claim cases elsewhere in the Pacific and participated in a week of advisory committee discussion on the Marshallese value of land. In 1999 and 2001, very helpful suggestions on methodologies, supporting research, and peer-review comments were provided by Marie Boutté, Michael Cernea, Norman Chance, Muriel Crespi, Susan Dawson, Ted Downing, Shirley Fiske, Judith Fitzpatrick, Jane Hill, Bill Johnston, Matthew Johnston, Terry Johnston, Robert Hitchcock, Bob Kiste, Ed Liebow, Gary Madsen, Bonnie McCay, Laura Nader, Theresa Satterfield, Ted Scudder, Anthony Oliver Smith, Amy Wolfe, and John Young. Our approach to working with Marshallese experts within the tribunal proceedings and the development of a November 2001 post-hearing brief on anthropologists as expert witness in land-claim proceedings occurred with substantive input from Paul Magnarella and with suggestions from Carmen Burch, Ted Downing, Kreg Ettenger, Richard Ford, Bill Johnston, Jane Hill, Bonnie McCay, Jerry Moles, Triloky Pandey, Wayne Suttles, and John Young. The approach to reparations was drawn in large part from the *Reparations and Right to Remedy* brief prepared by Johnston in 2000 for the World Commission on Dams and reflects the critical review comments of Ted Scudder, Robert Hitchcock, Dana Clark, Aviva Imhof, and Monti Aguirre. Our understanding of the relationship between the Marshallese research and the human radiation experimentation sponsored by the Atomic Energy Commission elsewhere (Japan, the Amazon, and the Arctic) was influenced by research suggestions and critical feedback from Terrence Turner, Norman Chance, and Louise Lamphere.[22]

Expert witness report production occurred with the assistance of Ted Edwards, Benjamin Edwards, and Christopher Edwards. In this latest effort to present *The Rongelap Report,* we gratefully acknowledge the production assistance of Kay Hagan and Hannah Shoenthal-Muse and the expert copyediting attention provided by Peg Goldstein. We thank Martin Sherwin for his review of an earlier draft and his enthusiastic support for the idea of publishing this expert witness report. And we thank Jennifer Collier, our Left Coast Press editor, whose suggestions markedly improved our efforts to frame the report.

---

[22] The broader history of Atomic Energy Commission–sponsored research involving indigenous populations in the Arctic, Amazon, Andes, Pacific, and American Southwest is discussed by Barbara Rose Johnston in the chapter "'More Like Us Than Mice': Radiation Experiments with Indigenous Populations," in *Half-lives and Half-truths: Confronting the Radioactive Legacies of the Cold War,* ed. Barbara Rose Johnston (Santa Fe, NM: SAR Press, 2007), 25–54.

Finally, to Bill Graham, whose critical attention to detail, helpful suggestions, editorial expertise, and unflagging commitment to tell the whole story demonstrated to us the many meanings of the title Public Advocate, we offer our deepest thanks and appreciation.

The Rongelap Report
Hardships and Consequential Damages from Radioactive Contamination, Denied Use, Exile, and Human Subject Experimentation Experienced by the People of Rongelap, Rongerik, and Ailinginae Atolls

Prepared at the request of:
Bill Graham, Public Advocate
Nuclear Claims Tribunal
P.O. Box 702
Majuro, Republic of the Marshall Islands 96960

Submitted by:
Barbara Rose Johnston, PhD
Senior Research Fellow
Center for Political Ecology
P.O. Box 8467
Santa Cruz, CA 95061
and
Holly M. Barker, PhD
Senior Advisor to the Ambassador
Republic of the Marshall Islands
Washington, DC 20008

September 17, 2001

CONTENTS

Part 5: Conclusions and Recommendations
Violations of Trustee Relationships
Statements of Culpability
Reparations
Relevant Case Precedents
Recommendations for Categories of Concern in This Claim
Concluding Remarks

Part I

Introduction

The U.S. nuclear testing program was conducted in the Marshall Islands from 1946 through 1958. The U.S. government detonated atomic and thermonuclear weapons with the aim of achieving world peace through a deterrence policy. This report demonstrates some of the ways in which the Marshallese people subsidized this nuclear détente with their lands, health, lives, and future. The evidence summarized within illustrates some of the consequences of the U.S. nuclear testing program. Its actions essentially inflicted nuclear war conditions on a fragile atoll ecosystem and vulnerable population. The Marshallese, despite appeals to the United Nations, were powerless to stop the testing and unprepared to address the proliferation of problems resulting from the testing.

This report presents findings from a collaborative, participatory, environmental anthropology research project exploring the human environmental impacts of nuclear testing, contamination, and exile as experienced by the people of Rongelap, Ailinginae, and Rongerik atolls.[1] Data from ethnographic research and a review of the scientific literature support findings that nuclear testing in the Marshall Islands resulted in contamination, short- and long-term exposure to radioactive substances, and alienation from land and other critical resources. Nuclear testing destroyed the physical means to sustain and reproduce a self-sufficient way of life for the people of Rongelap. Radioactive contamination and involuntary relocation radically altered health, subsistence strategies, sociopolitical organization, and

---

[1] In this report, "the people of Rongelap," or "Rongelapese," refers to those people exiled from Rongelap, Rongerik, and Ailinginae atolls. They are typically considered by other Marshallese as a single sociopolitical unit and are represented in the RMI government by one elected senator, although members of the Rongelap community retain rights to three atolls. Ethnographic interviews and historic records confirm the presence of villages and year-round residential use in each of the three atolls.

community integrity. A lifetime of service as human subjects in a wide range of biomedical experiments further harmed the health and psychosocial well-being of the people of Rongelap.

This report argues that the full extent of damage and injury, and the consequences of this damage and injury, must be considered when attempting to shape compensatory and remedial responses. Compensatory actions must reflect individual injuries and experiences, including pain, suffering, and hardship. Compensatory actions must also reflect the corporate experience of the people of Rongelap, whose health, vitality, and way of life have been fundamentally altered by the U.S. nuclear weapons testing program. Evidence presented in this report supports claims for compensation for:

- Social, cultural, economic, and political hardships and injuries experienced by the population of Rongelap as a result of the loss—through involuntary relocation and extensive contamination of terrestrial and marine resources—of the material basis for sustaining a healthy, self-sufficient way of life
- Psychosocial stigmatization, pain, and suffering experienced by the people of Rongelap as a result of their acute and long-term exposures to fallout
- Pain and suffering endured by members of the Rongelap community as a result of their involvement in long-term studies on the effects of radiation and their use as human subjects in a range of isolated experiments that had nothing to do with their individual health and treatment needs
- Natural resource damages and socioeconomic stigmatization experienced by the people of Rongelap, and the broader nation, as a result of contamination produced by the U.S. nuclear weapons program.

Summary of Relevant Findings

- The people of Rongelap experienced involuntary displacement from Rongelap and Ailinginae atolls when they were physically removed from their atolls (March–May 1946; 1954–1957). When people returned to their atolls, they lost access to a viable healthy ecosystem (thus they were displaced from their ability and rights to safely live in their environment in the years between 1957 and 1985). They became exiles (1985–present) when they were finally informed of the life-threatening contamination levels in their homeland.
- Families were deprived of their right to live and use lands on Rongerik Atoll.[2] Rongerik Atoll was taken for U.S. naval use following World War II and used

---

[2] Rongerik Atoll is the rare example of *morjinkōt* land—land that an iroij gives to a warrior for heroics in battle. The story of how this land became *morjinkōt* in the mid-1800s is still told today. See map 4.

as a weather and fallout tracking station during the nuclear testing program (1946–1958). The U.S. Navy, without getting permission or providing compensation, used Rongerik as a resettlement site for the Bikinians (1946–1948). And in 1957, when the Rongelap community was resettled on Rongelap and Ailinginae, the United States prohibited all access to and subsistence use of Rongerik Atoll due to severe contamination from nuclear weapons fallout.

- Exposure concerns involve much more than exposure to radiation and fallout from a singular testing event in 1954. Exposure concerns involve the persistent presence of contamination from sixty-seven atmospheric tests of nuclear weapons in the Marshall Islands. This contamination includes radioactive elements released·through nuclear explosions, as well as tracer chemicals, such as arsenic, used to "fingerprint" the fallout from each weapon. The people of Rongelap, Rongerik, and Ailinginae were exposed to external radiation and other toxic substances not only from fallout but more significantly from internal ingestion—breathing dust and smoke from household and garden fires, drinking water, consuming terrestrial and marine food sources, and living in houses and using material culture fashioned from contaminated materials.

- "Exposed" people of Rongelap include those living on Rongelap and Ailinginae in 1954 who were exposed to Bravo and other test fallout; those who were resettled in 1957; those who were born on the contaminated atoll; those who were exposed to materials and food originating from Rongelap, Rongerik, and Ailinginae atolls; and the descendents of people exposed to radioactive contaminants. Given the synergistic, cumulative, and genetic effects of long-term exposure to radioactive isotopes and other environmental contamination from military testing, exposure is of concern to this and future generations.

- The people of Rongelap, Rongerik, and Ailinginae, with other Marshallese, served as unwitting subjects in a series of experiments designed to take advantage of the research opportunities accompanying exposure of a distinct human population to radiation. The Atomic Energy Commission (AEC) initially funded human subject research involving the Marshallese in 1951 in an effort to document "spontaneous mutation rates" to better estimate the genetic effects of radiation produced through the nuclear weapons testing program. Research on the human effects of radiation was intensively conducted beginning in March 1954, with efforts to document the physiological symptoms of U.S. servicemen and Marshallese natives exposed to fallout from the Bravo test. Initial findings from this and other biological research projects helped shape the goals and approach of an integrated long-term study on the human and environmental effects of nuclear weapons fallout. That study began in 1954 and was continued by Brookhaven National Laboratory (BNL) through 1998.

- The people of Rongelap believe, and the documentary record confirms, that the United States was aware of the extraordinary levels of fallout from Bravo and subsequent tests, was aware of continuing levels of radioactivity, was aware of contamination in the marine as well as terrestrial ecosystem, was aware of the bioaccumulative nature of contamination, noted radiation-induced changes in vegetative and marine life that islanders relied upon for food, monitored the increased radiation burdens of the resettled people returned to Rongelap in 1957, and documented the human health consequences of this systematic and cumulative exposure. Medical exams, especially from the 1950s to the early 1970s, involved monitoring and diagnostic procedures meant to document bioaccumulation processes and physiological symptoms related to radiation exposure, rather than clinical efforts to treat the various radiogenic and related health problems of the people of Rongelap. Periodic "medical surveys" also subjected the people of Rongelap to procedures that produced biological samples—blood, marrow, teeth, and other samples were harvested and sent back to the United States—in support of a wide range of experiments, many of which had little or no connection to the individual health and treatment needs of the people of Rongelap. Varied human subject experimentation also occurred during medical treatment trips to research laboratories in the United States. Ethnographic and documentary evidence demonstrates that the experiences of human subjects were painful, abusive, and traumatic.

• In addition to biophysical injuries, exposure to the environmental hazards generated by the U.S. nuclear testing program (and related biomedical research) resulted in stigmatization and other psychosocial injuries that adversely affected individuals, the community, and the nation. Nuclear testing introduced new taboos: certain lands and foods were off-limits; marriage to certain people involved new social stigmas; birthing presented new fears and health risks; family life often involved the psychological, social, and economic burden of caring for the chronically ill and disabled. The failure of the U.S. government to provide the people of Rongelap with accurate information concerning environmental hazards and risks, coupled with contradictory pronouncements on what was and was not safe, created taboos that were incomprehensible yet dominated living conditions after the onset of testing in the Marshall Islands. This transformation in the loci of control over taboos from a Marshallese cultural realm to a U.S. scientific realm undermined rules and the customary power structures that shaped, interpreted, and reproduced strategies for living in the Marshall Islands. The fear of nuclear contamination and the personal health and intergenerational effects from exposure colored all aspects of social, cultural, economic, and psychological well-being. This imposed stigmatization adversely affected the economy, society, families, and individual health and well-being of the people of Rongelap, Ailinginae, and Rongerik, and to varying degrees the entire nation.

• After leaving Rongelap and Ailinginae atolls in 1985, the Rongelap community faced severe hardships as it struggled to rebuild some semblance of community in Mejatto, Majuro, Ebeye, and other locations. Involuntary resettlement placed hundreds of people on small bits of rented land, creating extremely dense, unsanitary, and impoverished communities. The Rongelap community represents some 8 percent of the total Marshall Islands population, and while an estimated 48 percent of the nation is able to support household needs through agricultural production, the Rongelap community is alienated from land and the traditional resources needed for survival. This loss of access affects diet, health, and household economy. Extremely dense residential patterns created by communities of exiles living on rented land have created or exacerbated terrestrial and marine pollution. The impoverished condition of the Rongelap community has intensified local resource use, and the ecosystemic viability of host island environs has been degraded. Restricted access to critical resources inhibits people's ability to teach younger generations how to sustain a self-sufficient way of life. Loss of access to customary lands further inhibits efforts to transmit key information across the generations—knowledge that is essential to the survival of the community if it is ever to return to customary lands. These social, cultural, economic, and environmental problems of urbanization are linked to the contamination and loss of lands and represent consequential damages directly linked to nuclear testing.

- Nuclear testing destroyed the means to sustain a self-sufficient way of life for the people of Rongelap. Customary uses of Rongelap, Rongerik, and Ailinginae atolls encompassed a rich range of social, cultural, and economic activities, values, and meanings that allowed for a vibrant, marine-based, self-sufficient, and sustainable way of life. Current and customary laws, traditions, and subsistence production patterns involve an inherited system of rights to both terrestrial and marine resources. The consequential damages of contamination from the nuclear weapons test program affect both terrestrial and marine ecosystems, including the natural and cultural resources that sustain life. Thus damage assessments and compensatory actions need to include consideration of lagoons, reef heads, clam beds, reef fisheries, and turtle and bird nesting grounds. Damage assessments and compensatory actions also need to address the loss of those resources important for sustaining the social and cultural aspects of life, including family cemeteries, burial sites of *iroij*, sacred sites and sanctuaries, and *morjinkōt* land. Because sustainable subsistence production, without overusing the resource base, requires access to multiple locations, cleanup and resettlement of the main island of Rongelap Atoll is not sustainable without restoration of all Rongelap's islands, as well as Rongerik and Ailinginae atolls. Compensatory actions should reflect a commitment to replace, restore, or create new means to sustain a self-sufficient way of life.

- Nuclear testing created environmental hazards, health problems, hardships, and other consequential damages that will persist for decades to come. Compensatory

actions should incorporate principles of nuclear stewardship and provide sufficient funds, facilities, expertise, and training to allow the people of the Marshall
Islands to conduct their own intergenerational epidemiological surveys and
environmental risk assessments, develop culturally appropriate environmental
risk-management strategies (including monitoring of contamination levels and
remediating terrestrial and marine ecosystems), and provide intergenerational
medical care.

Research Concerns

This expert witness report is the result of a research project initiated at the request
of Bill Graham, Public Advocate for the Marshall Islands Nuclear Claims Tribunal.
Graham sought anthropological input on culturally appropriate strategies for
assessing the value of land in a nonmarket environment for the purpose of awarding
just compensation for damages incurred on Rongelap, Rongerik, and Ailinginae
atolls as a result of the U.S. nuclear weapons testing program, and assistance in
identifying the consequential damages and losses experienced by the people of
Rongelap as a result of the program. Research was conducted by Barbara Rose
Johnston and Holly M. Barker,[3] beginning in November 1998, following Graham's
request for advice on how to value land from a Marshallese perspective. We began
with a review of appraisal documents prepared for the Bikini and Enewetak land
claims to consider the difference between an appraisal-based definition of land
value and Marshallese interpretations of the value, meaning, and significance of
land. We identified strategies for broadening the assessment and valuation process
in ways that reflected a Marshallese way of life. Because previous research suggested that the people of Rongelap suffered damages when they were physically
removed from their lands, when certain areas or atolls were declared off-limits,

---

[3] Barbara Rose Johnston is an environmental anthropologist and a senior research fellow at the
Center for Political Ecology in Santa Cruz, California. She is the former chair of the American
Anthropological Association Committee for Human Rights (1998–2000) and former director of
the Society for Applied Anthropology's Environmental Anthropology Project (1996–2000). Her
significant publications at the time of this report include the edited volumes *Who Pays the Price?
The Sociocultural Context of Environmental Crisis* (Island Press, 1994), *Life and Death Matters: Human
Rights and the Environment at the End of the Millennium* (AltaMira Press, 1997), and, with John
Donahue, *Water, Culture, Power: Local Struggles in a Global Context* (Island Press, 1998). From 1988
to 1990, Holly Barker lived in Mili Atoll in the RMI, where she served as an English instructor and
teacher trainer with the Peace Corps. Since 1990 she has divided her time between the Marshall
Islands and the United States, serving as a senior advisor to the ambassador for the Republic of
the Marshall Islands on political, health, environmental, education, and radiation-related issues.
Barker's PhD dissertation analyzed ethnographic and archival data related to the U.S. nuclear
weapons testing program conducted in the Marshall Islands from 1946 to 1958, with particular
emphasis on language used by communities in the Marshall Islands to convey their experiences
with radiation.

and when they were removed from the safe use of their lands and resources due to contamination from fallout, we identified the time frame for this research as extending from pre-testing through the present.[4]

An initial review found that in the Marshall Islands, rights to use critical resources are socially constructed, typically determined according to maternal relationships, legitimized according to communal relationships, and fluid rather than fixed. Existing appraisal documents considered one aspect of land valuation: the economic value of rights to use (lease) land. To consider damage and loss from a Marshallese perspective, we identified a number of key questions: What does land mean to the Marshallese? Can you assign a value to land? Does a valuation that assumes "property" is terrestrial space—the dirt underfoot—adequately incorporate all that a *weto* (a parcel of land extending from the ocean to and into the lagoon) represents? How do you value damage to the ecosystem and loss of a wide variety of natural resources (marine, terrestrial, arboreal)? How do you value the loss of access to and use of atoll resources and the loss of meaningful interaction between people and their environment? How do you value the loss of a self-sufficient way of life? What sort of valuation strategies might be used to articulate the broader range of damages and loss suggested by a human-environmental impact analysis of nuclear weapons testing in Rongelap, Rongerik, and Ailinginae atolls? These questions and considerations provided a starting point for developing an approach to assessing the value of land from a Marshallese perspective and suggesting some of the damages incurred by the Rongelap community as a result of nuclear weapons testing.

Research concerns were also shaped with reference to the Rongelapese experience as human subjects in ecological and medical research. For decades, the people of Rongelap have complained about medical and scientific researchers who document radiation-induced decay in their bodies and ecosystems yet ignore or fail to treat their broader health needs.[5] The Rongelapese also complain that research conducted by outsiders does not involve them and typically fails to consider the knowledge they have gained from their experiences with radiation. The Rongelapese serve as study subjects but rarely receive the results of research or recognize any direct benefit from research. Thus we developed a research approach that was purposefully transparent, involving people from Rongelap, Rongerik, and Ailinginae in shaping research

---

[4] See Holly M. Barker, "Fighting Back: Justice, the Marshall Islands, and Neglected Radiation Communities," in *Life and Death Matters: Human Rights and the Environment at the End of the Millennium*, ed. Barbara Rose Johnston (London: AltaMira Press, 1997); Barbara Rose Johnston, "Experimentation on Human Subjects: Nuclear Weapons Testing and Human Rights Abuse," in *Who Pays the Price? The Sociocultural Context of Environmental Crisis*, ed. Barbara Rose Johnston (Washington, DC: Island Press, 1994).

[5] See Robert A. Conard, letter to Dr. Charles L. Dunham, director of biology and medicine for the Atomic Energy Commission, regarding continued exposure of the Rongelapese, June 5, 1958. DOE archive, http://worf.eh.doe.gov/data/ihp1d/400212e.pdf (accessed April 1, 2008).

questions, suggesting knowledgeable informants to be interviewed, conducting research, analyzing findings, and refining recommendations. The Marshallese Land
Value Advisory Committee and a number of Marshallese informants reviewed an
earlier report on the value of land to the Rongelapese to ensure that translations
and ethnographic interpretations accurately reflected Marshallese perceptions and
experiences. The senator and mayor of Rongelap, key informants, and other experts
reviewed a draft of this document to ensure that the contents accurately reflected
the consequential damages and injuries suffered by the Rongelapese as a result of
the testing program. The research questions, strategies, and data-collection efforts
were supported by the collaborative involvement of Public Advocate Bill Graham,
Associate Public Advocate Tieta Thomas, and Marshallese anthropologist Tina
Stege. Access to supportive data was enhanced by the contributions of a peer-review
network of anthropologists, economists, sociologists, health physicists, lawyers,
and other environmental scientists.

Research Methods

Ethnographic sampling procedures used in this research included saturation
sampling and interviewing to sufficient redundancy, and snowball sampling. Data-
gathering techniques included spatial mapping, focus group interviews, and rapid
ethnographic assessment techniques. Rapid ethnographic assessments included
narrow, problem-oriented definition; participatory, rapid sampling of representative
sectors; small sample sizes; focus on cultural patterning and differences across
sectors of populations (rather than intercultural complexity or range of variation); a
systems perspective (with an emphasis on collecting information from all relevant
sectors in the community); cognitive techniques used to identify and assess
cultural domains; and relatively minimal use of quantitative sampling or survey
techniques.[6]

In March 1999, we traveled to Majuro and began working with the Marshallese
Land Value Advisory Committee to help shape fieldwork questions, plan research
strategies, identify knowledgeable informants, and generate public interest and
support for the project. The Land Value Advisory Committee was organized by the
Office of the Public Advocate and included members knowledgeable in customary
and current Marshallese resource relations (fishing; cultivation; production of

---

[6] As described in Robert Trotter and Jean Schensul, "Methods in Applied Anthropology," in
Handbook of Methods in Cultural Anthropology, ed. H. Russell Bernard (Walnut Creek, CA: AltaMira
Press, 1998), 691–736. We were also inspired by Richard Stoffle's use of ethnographic methods to
define and assess environmental risk associated with low-level radioactive storage. See R. Stoffle,
M. Traugott, J. Stone, P. McIntyre, F. Jensen, C. Davidson, "Risk Perception Mapping: Using
Ethnography to Define the Locally Affected Population for a Low-Level Radioactive Waste Storage
Facility in Michigan," *American Anthropologist* 93, no. 3 (1991): 611–35.

baskets, maps, and other material goods; medicinal ethnobotany; spiritual and other cultural uses and meanings of the landscape) and men and women who had lived on Rongelap and Ailinginae during the Bravo event and in the years following the 1957 resettlement. The Land Value Advisory Committee included Iroij Mike Kabua, Senator Wilfred Kendall, Councilwoman Lijon Eknilang, Councilman George Anjain, and handicraft businesswoman Mary Lanwi.[7] The advisory committee helped shape strategies to work with the Rongelapese to:[8]

- Establish the key variables that sustained their previously self-sufficient way of life by documenting traditional patterns of resource value, access, use, and control
- Identify key events and conditions that adversely impacted these resource relations and thus altered or destroyed people's ability to be self-sufficient
- Document their perceptions of the broader social and cultural damages associated with these damages and losses and assess some of the socioeconomic consequences of these changes.

In addition to a formal advisory committee, this research profited from the contributions of key members of the Rongelap community, representatives of the Rongelap local government, and representatives of the national government of the Republic of the Marshall Islands (RMI), including educators, environmental and health professionals, and elected officials and staff in the various ministries, who met with project consultants and provided background information and supporting documentation. The Majuro-based Rongelap community provided broad-based input during and following a public meeting in March 1999. This meeting introduced project goals and activities and resulted in comments, support, testimonials, and further involvement from Rongelap community members. A public meeting held in July 1999 gave community members the opportunity to review and discuss preliminary findings.

Holly Barker interviewed informants to compile testimony, often with the collaboration of Marshallese sociologist Tina Stege and Assistant Public Advocate Tieta Thomas. Testimony from Barker's 1994 research on the health consequences of nuclear testing (for the RMI government in support of a "changed circumstances" petition) was used to develop the chain of events and describe some of the

---

[7] We were deeply saddened by the death of George Anjain in February 2000. As mentioned later in this report, the Rongelap people believe that George's death, like many others, was a result of inadequate health care services.

[8] The political-ecological framework is historical and processual and allows analysis of the complex forces that stimulate and shape human environmental change. The emphasis on political and economic forces, ecosystemic conditions and dynamics, and sociocultural relationships is especially helpful in analyzing the social and cultural consequences of environmental crisis.

Rongelapese experiences. Interviews with the Rongelap community on Majuro, Ebeye, and Mejatto took place in March and April 1999. Follow-up interviews took place on Majuro in July 1999 and April 2001 and on Hawaii in July 2001. Barker and Stege also conducted interviews with the Rongelap community on Rongelap Island in 1999, during the first community visit to Rongelap since the 1986 evacuation. All interviews were conducted in Marshallese and later translated by Barker, with assistance from Stege. Barker speaks Marshallese, has more than a decade of experience and relationships in the Marshallese community, and has previously worked with many of the informants to record individual health problems associated with nuclear testing. Use of the Marshallese language ensured that interviewees could express themselves effectively. It also provided linguistic evidence of people's perceptions and experiences.

Access to Rongelapese informants was significantly enhanced by the efforts of Rongelap senator Abacca Anjain-Maddison and Rongelap mayor James Matayoshi, who provided introductions and encouraged the testimony of many members of the Rongelapese community. In June 2001, the senator helped arrange group interviews with members of the community who were in Hawaii for a meeting with the U.S. Department of Energy (DOE). The senator was present during most of the group discussions, assisted with the translation of traditional, older words, and encouraged people to share their experiences. More than seventy Rongelapese participated as informants in the formal interview phases of research.

Preliminary research findings on the value of land and damages associated with the loss of a way of life were initially drafted in July 1999 and submitted to the advisory committee, which met in Majuro to refine ethnographic interpretations and help develop research findings and implications. This meeting produced significant insights into compensatory concerns and recommendations. A revised report was submitted to Public Advocate Graham in October 1999.[9]

Between 1999 and 2001, Johnston conducted research on reparations and the right to remedy—identifying valuation procedures and precedent-setting court cases and settlements providing compensation for damages and hardships associated with involuntary resettlement, loss of a way of life, and human subject experiments. In March 2001, Johnston reviewed the Atomic Bomb Casualty Commission files at

---

[9] In 1999 anthropologist Stuart Kirsch attended the initial advisory committee meetings in Majuro, participating in discussions on the Marshallese value of land. He also prepared a briefing paper on valuation strategies used in other cases where claimants sought compensation for loss of a way of life. See Stuart Kirsch, "Appendix C: Case specific examples of compensatory action in the Pacific region," in Barbara Rose Johnston and Holly M Barker, *Assessing the Human Environmental Impact of Damage from Radioactive Contamination, Denied Use, and Exile for the People of Rongelap, Rongerik and Ailinginae Atolls: Anthropological Assistance to the Rongelap Land Valuation/Property Damage Claim* (Majuro, RMI: Nuclear Claims Tribunal Office of the Public Advocate, 1999).

the National Academy of Sciences archives in Washington, D.C. She identified documents describing funded research on the genetic effects of radiation beginning in 1951 and involving the Marshallese population. In April 2001, Graham asked us to rework our October 1999 draft report to add consideration of the hardships endured and consequential damages experienced by the people of Rongelap, and to outline appropriate remedies. Johnston reexamined the declassified literature to identify key documents demonstrating use of the Rongelap population in a range of human subject experiments. These findings were coupled with Barker's earlier research on the declassified documents. Barker collected additional testimony from Rongelap informants in Majuro (April 2001) and Hawaii (June 2001) concerning the hardships associated with involuntary resettlement and the pain and suffering experienced during human subject experiments. Interviews with members of the Rongelap community living in Hawaii added an important element to this research, highlighting the difficulties of managing radiation-related illness and gaining access to medical care as required by the 177 Agreement of the Compact of Free Association.

Throughout the history of the nuclear weapons testing program and related research on the human-environmental effects of nuclear weapons use in the Marshall Islands, information has been closely controlled by the U.S. government. Some of the secrecy has involved levels and content of fallout, fallout patterns, and the long-term bioaccumulative effects of fallout. The effects of radiation on humans, especially information relating to the wide range of studies and experiments involving the Marshallese people, has been the other major category of classified information.[10] The Clinton administration's declassification order in 1994 resulted in the disclosure of documents demonstrating the use of Marshallese subjects in a wide range of experiments, including purposeful intent to involve the population of Rongelap in long-term human-environmental studies on the hazards of fallout. Declassification and disclosure has not been exhaustive, however, and some of the material released includes significant deletions. The review of documents and the declassification of material continue. Because the declassification order relates to only those experiments involving the use of radioactive materials on human subjects, other experiments may still be classified. Thus this report reflects complaints about and requests compensation for abuses currently known and documented. We fully expect new information about further abuses to come to light.

---

[10] Advisory Committee on Human Radiation Experiments, *Final Report* (Washington, DC: U.S. Government Printing Office, 1995), 621–68. The ACHRE report is available online at http://www.hss.energy.gov/healthsafety/ohre/roadmap/achre/report.html. A record of ACHRE deliberations, including transcripts from committee meetings, interim reports, and a list of documents consulted, can be found at the National Security Archive: http://www.gwu.edu/~nsarchiv/radiation/.

Report Framework

The following report presents the findings from ethnographic and documentary research in ways that link anecdotal experience to the documentary record. Fifty years of human and environmental monitoring have produced substantive documentation of radioactive contamination, exposure, bioaccumulative hazards, and biodegenerative conditions. This documentary record is used to illustrate, substantiate, or punctuate Rongelapese testimony.

Testimony from the people of Rongelap was obtained through a series of interviews conducted with the understanding that comments and concerns would be recorded and used in this report and that this report would be submitted to the Republic of the Marshall Islands Nuclear Claims Tribunal and entered into the public record. Direct quotes from Rongelapese informants are included in this report as testimony. Their names, dates, and interview locations are also provided.

Names used in this report reflect spelling at the time of the interview or published account. As Marshallese is a fluid language with an evolving written tradition, many spellings vary.

Documents, publications, and other sources are mentioned in the text; full reference detail is provided in endnotes. U.S. government documents are referenced in three ways: Materials filed at the National Academy of Sciences Atomic Bomb Casualty Commission archives are referenced by file box title and number, document date, and title. Declassified materials posted on the DOE Web archives for human radiation experiments and the Marshall Islands are referenced by Web address. Materials presented to the RMI government by the DOE are referenced by author, document title, source agency, and archival repository. Documents presented by the U.S. government to the RMI were classified and not available to the Marshallese government or the general public during negotiations for the Compact of Free Association between the RMI and the United States. The majority of these documents were declassified in the mid- and late 1990s as part of the Clinton administration's inquiry into the use of human subjects in radiation experiments.

To fully identify and understand the consequential damages of the U.S. nuclear weapons testing program, it is important to consider the preexisting context, as well as the meaning and value of atoll resources from a Marshallese point of view. Thus part 2 of this document presents the physical ingredients, actions, and culturally constructed mechanisms that allowed the people of Rongelap, Rongerik, and Ailinginae to experience a self-sufficient, sustainable way of life. This summary of the key elements of the traditional subsistence system includes tangible resources, as well as the knowledge and social structures that allow effective, viable use of terrestrial and marine resources. In demonstrating the key elements of a viable system, this summary provides the ethnographic and environmental detail supporting research findings that contamination and involuntary relocation fundamentally altered, and in many instances destroyed, the material basis for sustaining a healthy, self-sufficient way of life; consequential damages include damage to and loss of natural and cultural resources and the means to produce and reproduce a sustainable way of life.

An annotated outline of the chain of events and related concerns associated with the U.S. nuclear weapons testing program and its human subject research program is presented in part 3. Key events include the contamination of an entire population and their homelands, removal of people from the safe use of their surroundings, injuries that produced the need for lifelong and intergenerational medical treatment, and the use of human subjects in scientific research programs. Informant testimonies concerning these events have been included to further contextualize experiences and to suggest some of the psychosocial, cultural, economic, and political hardships and injuries experienced by the population of Rongelap as a result of exposure to fallout, medical research, extensive contamination of terrestrial and marine resources, and involuntary relocation.

Part 4 presents a summary of damages from these events as experienced by the people of Rongelap and the broader nation. The consequences of these events include economic, political, social, cultural, and psychological injuries and hardships experienced by individuals, households, kin networks, the broader corporate group, and the nation as a whole. This section includes the pressing questions, comments, and remedial suggestions offered by Rongelapese informants, and those emerging from an anthropological analysis of events and experiences.

Part 5 addresses the question of meaningful remedy from a broader legal perspective by identifying and defending major categories of damage and loss, reviewing significant statements of culpability, and citing relevant valuation tools, compensatory actions, court findings, and settlement precedents in similar cases in the United States and internationally.

▲ 1 Looking down on Majuro, April 1999. The Marshallese live in a water world—on small, narrow islands of sand that lie atop coral reefs. The reefs form a chain of islands, an atoll, that frames a large lagoon. In a nation the size of Mexico, stretching across some 750,00 square miles of ocean with 870 reef systems and 1,225 islands, the total landmass amounts to some 70 square miles. While land, for the Marshallese, is precious and valuable, the notion that a monetary value can be assigned to land, as well as the concept that people can own land, suggests a superficial understanding of the world and how it works. In the Marshall Islands, land is the physical framework for both a material realm that supports and sustains life, and a spiritual realm that brings meaning, cohesion, and happiness to people's lives.
Photo by Barbara Rose Johnston

◂ **2** Marshallese sailing canoes, 1899. Marshallese are renowned across the Pacific for their navigational skills. The nation is spread out over huge distances and it took many days to sail between each atoll, yet the people were united in a common language and culture. Travel by sail was regular, common, and an essential element of Marshallese life.
Courtesy Pacific Trust Territory Archives

◂ **3** Man in an outrigger canoe on Rongelap Lagoon, early 1950s. Wood is scarce in the Marshall Islands and sailing vessels, large and small, were cherished and used over several generations.
Courtesy Pacific Trust Territory Archives

▸ **4** "Rongelap Man and Two Women in Costume Outside Pole and Thatch Temporary Shelter; Pole and Thatch House in Background, 1899."
Photo by Henry Clifford Fasset, courtesy Smithsonian National Anthropological Archives

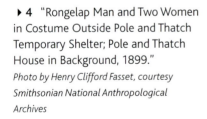

▸ **5** Copra drying, 1957. For the first half of the 20th century the people of Rongelap produced the bulk of their needs from local agriculture and fishing. Cash was generated through the occasional sale of copra to passing traders and used to buy rice, matches, and other small necessities.
Courtesy Pacific Trust Territory Archives

6 Able Test, 1946, demonstrating the power of atomic weapons. Ten congressmen, 2 senators, and 166 newsmen, including 10 representatives of the foreign press, attended the Able Test, along with some 42,000 members of the U.S. military. The United Nations Observer Group also attended, with official representatives of the member countries of the United National Atomic Energy Commission: Australia, Brazil, Canada, Chile, France, Egypt, Great Britain, Mexico, Netherlands, Poland, and the U.S.S.R.

U.S. Navy photo, courtesy National Archives

▲ **7** Baker test, 1946. This classic image appeared in newspapers around the world. The underwater detonation atomized the lagoon floor and sent tons of material into the atmosphere. When the column collapsed, some ten million tons of water fell back into the lagoon, blanketing the remaining ships with radioactive spray, mist, and debris. The first wave from the collapsed tower rose 94 feet high. The military conclusion from the 1946 tests: no ship within a mile of atmospheric or underwater detonation could escape without serious damage to the ship or crew.

U.S. Navy photo, courtesy National Archives Collection

▲ 8 George Test, Operation Greenhouse, May 8, 1951. This 225 kilotons bomb, detonated at Enewetak, was the first true test of thermonuclear fusion and a precursor to the hydrogen bomb.
U.S. Department of Energy

▲ 9 Operation Ivy's shot King, a weapons-related air drop on Enewetak on November 15, 1952. Some 195,000 U.S. service members have been identified as atomic veterans exposed to radiation as participants in the post–World War II occupation of Hiroshima and Nagasaki following the atomic bombing of Japan. In addition, approximately 210,000 individuals, mostly military peronnel, are confirmed as participants in U.S. atmospheric nuclear weapons tests between 1945 and 1962 in the United States and the Pacific and Atlantic oceans prior to the 1963 Limited Test Ban Treaty. Another estimated 90,000 servicemen were exposed to radiation from the underground tests in Nevada.

Courtesy of National Nuclear Security Administration / Nevada Site Office

▲ **10** Bravo Test, Operation Castle, March 1, 1954. The Bravo test was the first detonation of a hydrogen bomb dropped from an airplane. The largest detonation by the United States, Bravo had a radioactive cloud that plumed over 7,000 square miles. It was designed specifically to generate the largest possible cloud of fallout.
U.S. Department of Energy

▼ **11** "Project 4.1," documenting and beta burns (*left*) and loss of hair (*right*).
U.S. Navy photo, courtesy Republic of the Marshall Islands Embassy

 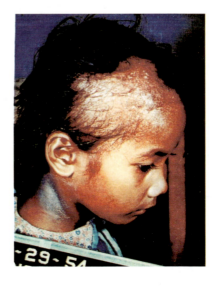

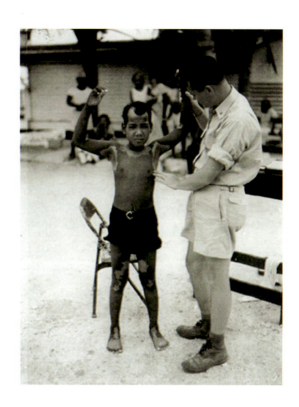

▲ 12 Examination of burns
suffered by a Rongelap boy.
*U.S. Navy photo, courtesy Republic of
the Marshall Islands Embassy*

▶ 13 Monitoring radiation
levels on Rongelap after
Bravo test.
*U.S. Navy photo, courtesy Republic of
the Marshall Islands Embassy*

▲ ▲ 14   Exhuming those who had died during the three years of exile following the Bravo test. These bodies were taken back to Rongelap for proper burial in 1957.
Courtesy Pacific Trust Territory Archives

▲ 15   Loading personal possessions on the Navy LST, return to Rongelap, 1957.
Courtesy Pacific Trust Territory Archives

◀ 16   On board, return to Rongelap, 1957.
Courtesy Pacific Trust Territory Archives

▲ **17** Rongelap children lining up for the medical survey exams, Rongelap clinic, 1961.
Courtesy Pacific Trust Territory Archives

▶ **18** Working with "control subjects": schoolchildren on Majuro, Marshall Islands.
Courtesy Pacific Trust Territory Archives

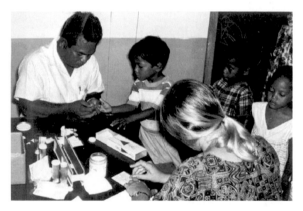

▼ ▶ **19** Dr. Robert Conard examining thyroids on Rongelap, circa 1973. Thyroid disease and cancer became a common condition on Rongelap. By December 2006, the Nuclear Claims Tribunal had issued some 1,186 personal injury awards to Marshallese with documented thyroid abnormalities. Today, in addition to other radiogenic disease, Marshall Islanders have one of the world's highest rates of abnormalities of the thyroid.
Courtesy Brookhaven National Laboratory

▲ **20** Evacuation of Rongelap, May 1985. This picture of Aisen Tima, the Rongelap school teacher, was taken through the porthole of Greenpeace Rainbow Warrior, with Rongelap in the background.
Courtesy Glenn Alcalay

◀ **21** Greenpeace assisted in the evacuation of Rongelap in 1985 after the island had been contaminated by fallout from atmospheric nuclear weapons tests.
Photo by Fernando Pereira, copyright Greenpeace

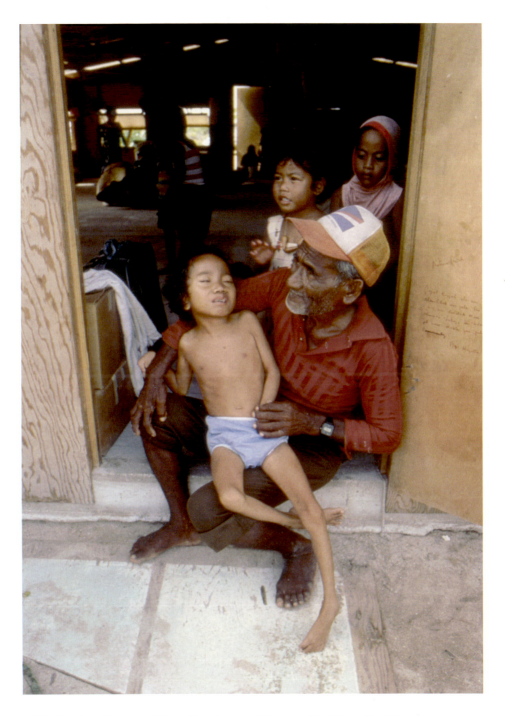

▲ 22 Carol and her grandfather. Carol is one of the severely deformed children born on Rongelap. Because Carol's mother was not on Rongelap during the Bravo test, but moved back to be with her family in 1957, she was considered "unexposed." Carol's congenital conditions are not an acknowledged and compensable radiation-related injury.

Photo by Fernando Pereira, copyright Greenpeace

Figure 18. Brothers. Marked retardation in statural growth is shown by the older (shorter) brother (No. 3, on the right) who was exposed at age 18 months. The younger by 21 months (No. 83, on the left) is taller by 13 cm. The retarded boy shows no evidence of hyperthyroidism or skeletal disease clinically, other than markedly delayed osseous maturation.

◄ **23** Brookhaven scientists document stunting. Much of the medical survey research in the first decade focused on the physical effects of radiation exposure as demonstrated in changes in the skeletal growth of children on Rongelap during the Bravo detonation, and comparing these growth rates with other Marshallese children. Many of the "control subjects" lived on Rongelap but were not considered "exposed." Both brothers later developed thyroid disease. *Figure 18 in Robert Conard et al, "Brookhaven report of the 1963 and 1964 medical survey" (BNL 908 [T-371]). Courtesy Bill Graham*

▼ **24a, b** Intergenerational worries. In 1982 journalist Giff Johnson studied possible nuclear-related health problems experienced by second- and third-generation children. In response to a request to take pictures of affected children living on Majuro, this child (below left) was brought to Giff by his parents. It is unclear if the child's parents were from Utrik, Rongelap or another atoll. The photo on right (photographer unkown)—apparently the same child—was used in the Nuclear Claims Tribunal hearing to illustrate hardships that Marshallese families endure when caring for children with congenital defects. This child is stunted with very short legs, a withered arm, and an overly large head.

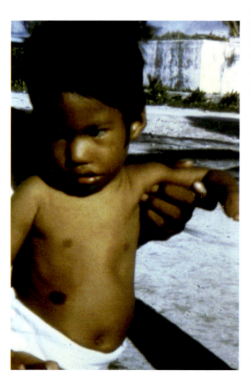

Erkein jet ian utamwe ko: jibikbik, im utamwe in kōmālij.

These are some defects: malformations and mental defects.

▲ **25** DOE assurances. This drawing depicts a boy with a withered arm and a girl with Down's syndrome. It was created by the DOE to illustrate a Marshallese/English language booklet published in response to concerns that the high incidence of miscarriage and birth defects in Rongelap and Utrik were due to living in a radioactive environment. In addition to the caption below the image, the text states: "There are some kinds of sicknesses or infirmities that can occur in any baby which it got from its mother or father," and: "Some babies die before they are born, and some are born with defects." The overall intent was to assure the Marshallese that these birth defects were normal and to be expected, had nothing to do with the Nuclear Weapons Testing fallout, and that their islands were safe places to live. While this document acknowledged that exposure to I-131 produced thyroid abnormalities, it did not acknowledge related health effects. Impaired thyroid function in pregnant women can result in cases of retardation, cretinism, and stunted development in their offspring.

"The Meaning of Radiation for Those Atolls in the Northern Part of the Marshall Islands That Were Surveyed in 1978" *(DOE/NBM—1052), p. 27, courtesy Bill Graham*

▲ ▲ 26   Return to Rongelap, 1999. In preparing the Rongelap claim, the Nuclear Claims Tribunal visited Rongelap with appraisers, scientists, and members of the exiled Rongelap community in April 1999. For the people of Rongelap, this was their first visit back to Rongelap since evacuation in 1985.
Photo by Holly Barker

▲ 27   Lijon Eknilang testifying to the RMI Nuclear Claims Tribunal, November 1, 2001.
NCT videotape

▲ ▲ **28** Marshallese march on Nuclear Victims/Survivors Day, March 1, 2004, Marshall Islands. Nuclear Remembrance Day is commemorated around the world on the anniversary of the Bravo Test, and on this 50th anniversary people wore Project 4.1 shirts to remember past suffering and to draw public attention to a largely unknown chapter of U.S. history: the use of the Marshallese as human subjects in the Project 4.1 radiation research.
Photo by Holly Barker

▲ **29** Ladies of Rongelap at a community gathering on Majuro. September 2004.
Photo by Holly Barker

▲ 30   Protest over cuts in health care funding. Rongelap Councilwoman Rokko Langinbelik at a bilateral meeting between the U.S. Department of Energy and the Republic of the Marshall Islands, Majuro, September 2004. Councilwoman Langinbelik and other members Rongelap and other exposed communities protested announced cuts in the level of medical funding provided by the U.S. Department of Energy for the radiation-affected population. *Marshall Islands Journal*

◀ 31   This image was presented on July 19, 2005 to the U.S. Senate Energy and Natural Resources Committee during hearings convened by Honorable Eni Faleomavaega (D-Samoa) regarding the health care needs of Marshallese exposed to radiation. Reportedly, there have been at least six children—all grandchildren of the original survivors—who were born on Utrik in recent years so badly deformed that each died within weeks. *Photo by Utrik Senator Hiroshi V. Yamaura*

Part 2

Loss of a Healthy, Sustainable Way of Life

This section presents evidence to suggest that consequential damages and injuries associated with the U.S. nuclear weapons testing program include the loss of the means to maintain a healthy, sustainable way of life. In its review of compensation claims for the people of Bikini and Enewetak, the Nuclear Claims Tribunal has awarded compensation for loss of land based on valuation of land derived from a record of leases in Majuro, Kwajalein, and other atolls. Such valuations reflect one aspect of economic value—the value of temporal occupation of dry acreage. Such valuations do not consider and adequately address social and cultural values, meanings, and uses of land. Nor do these valuations address the wider realm of arboreal and marine resources that sustain the Marshallese way of life. In this section we focus on the complex social, cultural, political, and economic relationships between the people of Rongelap and their atolls. We outline customary rules and obligations, identify critical resources, and describe some of the mechanisms used to care for and transmit knowledge about the land and the sea. This human environmental detail provides a framework for interpreting subsequent damages and injuries experienced by the people of Rongelap when removed from the healthy use of their homelands.

Valuing Land from a Marshallese Perspective

Outside observers traveling to the islands from their continental homes have described the Marshall Islands as a very small place—tiny spits of sand and reef forming islands and atolls and separated by immense stretches of open sea. In actuality, the Republic of the Marshall Islands extends across some 750,000 square miles of ocean, an area equivalent to the landmass of Mexico. Within this national boundary lie an estimated 870 reef systems and some 1,225 islands distributed among two island chains: the Ratak (sunrise) chain and the Ralik (sunset) chain. These islands form twenty-nine atolls and five individual islands, with a total estimated dry landmass of some 70 square miles. For an estimated four thousand years, the people of the Marshall Islands have survived and thrived on thin stretches of land in a universe of ocean. From their perspective, the world is largely water, and their terrestrial lands are part of—rather than separate from—the sea. Survival has required knowledge, experience, and access to critical resources in both the fixed and the fluid realm.

Descriptions of the land by outside observers often emphasize severe limitations: "Of all the island and atoll types in Micronesia, the drier coral atolls," such as Rongelap, Ailinginae, and Rongerik, "present the greatest challenge to human occupancy."[1] These low-lying, dry atolls have "slender resources" [and] "practically no soil covers the coral, and the inhospitable sand will grow few plants."[2]

Ethnographic interviews with the people of Rongelap, Ailinginae, and Rongerik paint a different picture. For example:

Lijon Eknilang: *The islands in the northern islands, like Rongelap, they were the garden islands. Everyone knew they were the best place to grow arrowroot, pandanus . . . there were whole islands covered with birds, and turtles.*[3]

Kajim Abija: *There was always pandanus and coconuts. During the times of hunger, we were not really hungry because there was so much of the pandanus.*[4]

Mwenadrik Kebenli: *I was born in 1917 on Rongerik. We came together and made food. We stored food. So many kinds of food. Every type was there. There was water. We ate arrowroot, preserved breadfruit, pandanus, coconut, fish, crab, foods from the reef like octopus, clam, and small fish. There's an island, Jipedbao, with lots of birds. There were also turtles on the islands. We would sail overnight in the outrigger canoe to Rongelap. We would take food to Rongelap. We always exchanged food. There were so many good things on Ailinginae.*[5]

Dorothy Amos: *There was a lot of coconut, lots of pandanus and foods we make from pandanus, also arrowroot. Foods from the ocean, there were so many. Big clams, crab, every kind in the depths of the ocean. We ate, or sold them to boats that would come in, we could exchange shells and food supplies. So many crabs, and birds. We got life from them. There*

[1] Leonard Mason, "The Ethnology of Micronesia," in *Peoples and Cultures of the Pacific,* ed. Andrew Vayda (Washington, DC: Natural History Press, 1968), 278.
[2] Herbert W. Krieger, *Island Peoples of the Western Pacific: Micronesia and Melanesia* (Washington, DC: Smithsonian Institution, 1943), 21.
[3] Lijon Eknilang, advisory committee meeting, Majuro, March 4, 1999.
[4] Kajim Abija, interview by Holly M. Barker, Ebeye, March 16, 1999.
[5] Mwenadrik Kebenli, interview by Holly M. Barker, Ebeye, March 18, 1999.

was good well water on Ailinginae. There was also good food on the small islands around Rongelap, not just on Ailinginae.[6]

These contrary perceptions of the resource base as limited or lush reflect different worldviews. From a continental perspective, limited land meant limited resources—minimal agricultural production; no rivers, lakes, or streams; relatively few trees. From a Pacific Islands point of view, land provided access to an immense array of resources. Cultivation and husbandry of trees and bushes were elements of a broader subsistence strategy that utilized the resources provided by the currents of wind and water; migratory birds, fish, and marine mammals were all essential elements of the Marshallese subsistence economy. Critical resources included not only terrestrial and marine materials and foods necessary for basic survival but also the knowledge about how to use and exploit resources in sustainable ways.

Land and Sea Tenure

The importance of land to the Marshallese, not only as a place to live and grow food but also as the essence of Marshallese life, is well documented. On April 20, 1954, the Marshallese citizens of the Trust Territory of the Pacific Islands submitted a petition to the United Nations Trusteeship Council. In this petition, the Marshallese people cite their concerns about the U.S. nuclear weapons testing program—concerns that included damage to health and the long-term implications of being removed from their land:

> *We, the Marshallese people feel that we must follow the dictates of our consciences to bring forth this urgent plea to the United Nations, which has pledged itself to safeguard the life, liberty and the general well being of the people of the Trust Territory, of which the Marshallese people are a part.*
>
> *. . . The Marshallese people are not only fearful of the danger to their persons from these deadly weapons in case of another miscalculation, but they are also very concerned for the increasing number of people who are being removed from their land.*
>
> *. . . Land means a great deal to the Marshallese. It means more than just a place where you can plant your food crops and build your houses; or a place where you can bury your dead. It is the very life of the people. Take away their land and their spirits go also.*[7]

---

[6] Dorothy Amos, interview by Holly M. Barker, Ebeye, March 18, 1999.

[7] United Nations Trusteeship Council, "Petition from the Marshallese People Concerning the Pacific Islands: Complaint Regarding Explosion of Lethal Weapons within Our Home Islands," April 20, 1954. The response to the Marshallese petition noted, "The Administering Authority adds that any Marshallese citizens who are removed as a result of test activities will be reestablished in their original habitat in such a way that no financial loss would be involved." United Nations Trusteeship Council, "Petitions Concerning the Trust Territory of the Pacific Islands," July 14, 1954, 5.

This paramount role of land rights in traditional Marshallese society was documented by Jack A. Tobin, district anthropologist for the U.S. territory of the Marshall Islands. An interviewee told him: "We will never willingly accept any other land in exchange for our lineage land." As Tobin noted, the concept of alienation of lands by sale, lease, or rental was "introduced by foreigners." Marshallese land tenure "forbids sale of land to non-indigenes," and the "Marshallese will not willingly accept complete alienation of their lands."[8]

The Marshallese value land as an integrated part of the marine ecosystem, a fact noted in Tobin's reports. In his 1958 report Tobin noted that property rights extend from terrestrial property into the marine area: "According to custom, the property rights extended out to the area where people stood, usually waist deep, in order to fish with a pole. These rights belonged exclusively to the lineage, whose land holding bordered the marine area."[9] Tobin also noted that the marine resources belonged to the iroij, but the people residing on an atoll were free to use the resources:

> *Traditionally everything of value in the lagoon such as shellfish, langusta, etc., was considered to be the property of the chiefs. The inhabitants of the particular atoll did not have to ask permission to take these items unless they were tabu property of the chiefs: the concept that the right to exploit the marine resources of an atoll is the prerogative of the inhabitants of that atoll only still persists.*[10]

Customary resource relations involve the rules and traditions governing access to and use of critical resources, as well as the meanings and values assigned to critical resources. The Marshallese are not alone in establishing rules and traditions to govern marine property. Anthropologist John Cordell, in his analysis of sea tenure systems, notes that throughout the Pacific, "seascapes are blanketed with history and imbued with names, myths, legends, and elaborate territories that sometimes become exclusive provinces partitioned with traditional rights and owners much like property on land." Cordell notes that evidence of marine property systems is reflected in the localized and specific knowledge of reefs, tides, currents, winds; in the traditions that limit access to and knowledge of fishing areas to those with specific entitlements; and in the names for territories, subsurface features, rocks, and reef clefts—names that represent events and mythical characters and provide local people with a "constant visible historical anchor."[11]

---

[8] Jack A. Tobin, *Land Tenure Patterns in the Trust Territory of the Pacific Islands* (Guam: Office of the High Commissioner of the Trust Territory of the Pacific Islands, 1958), 21, 1, 3.

[9] Tobin, *Land Tenure Patterns,* 57.

[10] Ibid.

[11] John Cordell, *A Sea of Small Boats* (Cambridge, MA: Cultural Survival, Inc., 1989), 1, 9.

Rules Governing Access and Use Rights

In his 1958 report on customary land tenure in the Marshall Islands, Tobin observed that "land, which is the source of their day-to-day existence, is considered by the Marshallese to be their most valuable asset,"[12] and while this asset is valued, it is not owned. According to Tobin, the Marshallese property rights system is "premised on the notion that no single person owns the land. Rather, the three-tier social structure acts like a corporation that collectively oversees the best interests of the property. The lineage (*bwij*) members may live on and exploit the resources of the land parcel or, if they possess rights in more than one land parcel, as is usually the case, merely make copra on it and use its food resources such as: coconuts, breadfruit, arrowroot, pandanus, bananas, and taro. Pigs and chickens are kept and fish and shellfish are obtained from the adjacent marine areas."[13]

Maintenance of the property and resources is critical to ensuring the continued livelihood of everyone in the corporation with rights to the property: "On certain occasions, landowners may allow non-landowners to live and work on their property. Members and associated members of the lineage (*bwij*) work the land, clearing it of underbrush and performing other tasks necessary for the simple type of agriculture practices in these low-lying coral atolls with their limited resources. In some instances people are allowed to work land not belonging to their lineage when members of another lineage have more than enough land for their own needs, or want to help some less fortunate person."[14]

In this type of nonfamily use arrangement, the nonlandowning tenant acts, in Western terms, as a renter. The renter is expected to provide a portion of cultivated resources or profits from the resources to the landholders.[15] Additionally, the renter is expected to maintain the land by using resources sustainably and ensuring the future production capacity of the property and resources. If the renter is not a responsible steward of the property and resources, the landholder will evict the renter. Landholders benefit from this rentlike arrangement by having their property and resources cared for responsibly. By the same token, the renter benefits by accessing resources necessary for survival. In addition to caring for property and resources for the well-being of the existing generation, people with land rights work collectively

---

[12] Tobin, *Land Tenure Patterns,* 2.

[13] Ibid., 7–8.

[14] Ibid., 10.

[15] Tobin describes land rights in terms of lineage ownership but uses the term *landowner.* The Marshallese have no word for landowner. In this report, we use the term *landholders* to refer to the current generation of people with recognized rights and responsibilities to the land. This use is at the request of the Land Value Advisory Committee, one of whose members (Lijon Eknilang) remarked on March 3, 1999: "Can we change the word *landowner*? We do not own the land, it belongs to the generations."

to ensure that lineage holdings will be productive for the succeeding generations that will inherit the land. In this regard, customary ownership of property and resources is based on sustainable interactions with the environment and responsible stewardship that allows future generations to flourish.

The national law of the Republic of the Marshall Islands reflects and codifies this customary law, protecting landholders and users and their customary traditions that allow sustainable subsistence in a fragile island environment. Section 5.1 of the Constitution of the Republic of the Marshall Islands recognizes that the "means to obtain subsistence and benefits is of utmost value. For this reason, the RMI Constitution upholds the notion of just compensation for all interest holders of the land when land is taken."[16] Because the Marshallese view themselves as intergenerational holders of the land, interest holders include future generations that will have rights to the land.

The excerpts from Tobin and the Constitution of the Republic of the Marshall Islands illustrate a critical feature of the Marshallese subsistence system: it is a tightly structured, common-property-rights system in which rights to use land and other critical resources are inherited. Our ethnographic research confirmed the presence and importance of land-use rights transferred between generations. The Western notion of individual property rights (meaning the ability to own, sell, or buy a piece of land), and thus a free market in land, does not exist. In exceptional cases, individuals with rights to wetos legally transfer their rights (to release themselves from serious debt, for example). However, even in these cases, transactions involve the transfer of use rights between Marshallese families, with the transactions regulated and controlled by customary authority. Non-Marshallese individuals cannot purchase land. Rights to use land are occasionally exchanged for cash payment. But customary and current use rights are highly prescriptive, typically involving the right to occupy, and do not include rights to freely use without permission coconuts, pandanus, and other material resources situated on the land, or reef heads and other marine resources adjoining the land. Customary exchanges of use rights, and many current use-rights agreements, typically involve the exchange of labor and the products of labor for the right to occupy and use, with permission, the natural resources of a particular weto or area. Traditional and current use-rights arrangements also include an implicit agreement to care for the land, including husbandry of trees, clearing and burning underbrush, and improving the fertility of soil. The right to occupy or use a weto can be revoked when stewardship principles are not being met.

Land rights in Marshallese society represent far more than the economic means to sustain a household:

---

[16] Constitution of the Republic of the Marshall Islands, section 5.5.

Mike Kabua: *Land gives you the meaning of life and the role of each individual in society.*[17]

Wilfred Kendall: *The people here have tenaciously held onto land. The resource people treasure most is land. Land speaks of your being, essence, reason for living. You relate to the world in terms of land [that] provides for your present, future, and future needs. . . . You cannot put enough value on land. . . . How do you put a value on something that people consider as a living thing that is part of your soul?*[18]

The pattern of access and use rights to property and natural resources reflects social relationships and standing in the community:

Dorothy Amos: *We have different customs than Americans. They won't say there are three tiers of land rights, but for us, there are. We protect the land. This is our inheritance from our grandfathers and grandmothers. We bury on our land. We do planting.*[19]

Customary traditions and current practices involve a fluid rather than fixed system of rights to wetos. Weto rights are inherited, with a woman often investing a male member of her family with the power to claim use rights. To claim the right to use a weto, one must be able to describe it in relation to other wetos. The power to recognize and validate that claim rests in the hands of customary authorities (*alabs,* or land managers, and iroij). Customary authorities adjudicate conflicts over family claims. Changes in family structures or customary power structures (for example, the death of a mother, an *alab,* or an iroij) can produce a redistribution of use rights. Monetary compensation for denied use or damage to land, based on the right to use land as defined and recognized at a given time, can in effect impose a system of individual property rights. This system has the potential to undermine the traditional balance of power between men and women.

Marshallese culture and society revolve around a three-tiered structure, consisting of the iroij, the *alabs,* and the *ri-jerbal* (those who work the land). Because the social structure of the Marshall Islands is extensively documented elsewhere,[20] this portion of our report focuses on Rongelapese perspectives about this three-tiered social structure as it relates to property and access to resource rights.

---

[17] Mike Kabua, advisory committee meeting, Majuro, March 4, 1999.
[18] Wilfred Kendall, advisory committee meeting, Majuro, March 3, 1999.
[19] Dorothy Amos, interview by Holly M. Barker, Ebeye, March 18, 1999.
[20] Jack A. Tobin, *The Bikini People, Past and Present* (Guam: Office of the High Commissioner of the Trust Territory of the Pacific Islands, 1953); Tobin, *Land Tenure Patterns;* Mason, "Ethnology of Micronesia"; Robert C. Kiste, *The Bikinians: A Study in Forced Migration* (Menlo Park, CA: Cummings Publishing Company, 1974).

Iroij

Although the iroij did not reside on Rongelap, Rongerik, or Ailinginae, everywhere in the Marshall Islands, the iroij is responsible for the people. He or she must provide for people in times of need and distribute resources so that everyone is taken care of. The iroij is the key to the customary resource-management system. It is his or her job to decide the rules of access and use; hear and settle disputes over fish weirs, stolen fish, breadfruits, and coconuts; and determine weto boundaries, rights to use wetos, and succession rights:

> **Mike Kabua**: *The iroij always protects their people like a fence wrapped around them, standing at the gate taking care of them.*[21]
>
> **Lijon Eknilang**: *The iroij is like our government.*[22]
>
> **Wilfred Kendall**: *The iroij have people clean the land. People get to use the resources on the land and the iroij benefit from having the land maintained.*[23]

Alabs

Alabs, or managers of the land, keep records about the land and the wetos. They provide daily oversight of the land and the workers. *Alabs* also carry the rights to their lineages—their "property."

> **Mike Jenkins**: *Alabs essentially act like a corporation. There is a system of checks and balances in place. You can't have private ownership anymore. Instead, the corporation looks after your trust, or jolet. The alab is the chairman of the board. The alab can't make an agreement that is binding on our next generation. It is important to understand the holding of the corporation (what land), the conditions of the land, and what decisions you are allowed to make regarding the land. Alabs are closer to the iroij than other people in the community because they bring them their food when they visit and they serve as intermediaries between the workers and the chief.*[24]

Ri-jerbal or Kajor

While the *alabs* control the day-to-day work on the land, the ri-jerbal has the rights to use the land. The word *ri-jerbal* developed during the Japanese occupation

---

[21] Mike Kabua, advisory committee meeting, Majuro, March 4, 1999.
[22] Lijon Eknilang, advisory committee meeting, Majuro, March 4, 1999.
[23] Wilfred Kendall, advisory committee meeting, Majuro, March 5, 1999.
[24] Mike Jenkins, interview by Holly M. Barker, Gugeegue, March 15, 1999.

(1914–1944) when people began to collect coconuts for commercial value.[25] *Ri*, literally "the people," and *jerbal*, or "work," defined those who did the work for money. *Kajor* is a traditional word used to describe the third tier of society. *Kajor* means "strength" in the Marshallese language. It reflects a symbiotic relationship in which the kajor provides strength or support to the *alabs* and iroij. While *alabs* control and manage the land, the ri-jerbal maintains the right to use the land and its resources necessary for survival. If the ri-jerbal does not take proper care of the property and its resources, the iroij might forbid the ri-jerbal from using it. The majority of the people of Rongelap, Rongerik, and Ailinginae are the ri-jerbal. They have the right to work and maintain the land in exchange for sustenance and survival from the land.

The Rongelapese recognize the iroij's ultimate authority over natural resources. As a demonstration of this authority, certain foods and areas of the reef and land are set aside for iroij:

Lijon Eknilang: *Certain species are used just for the iroij, such as brown eel, turtle, whales, dolphins, and frigate birds. Other coral heads in a lagoon are reserved just for the iroij as the place to collect food for the iroij. It is mo, or forbidden, to go to islands and areas reserved for the iroij.*[26]

Mitjua Jankwin: *When he [the iroij] was here [on Rongelap], we gave him food from those coral heads. Sometimes food from those coral heads was sent to Ebeye. We sent janwin (preserved breadfruit), too.*[27]

Beyond iroij authority over critical resources, all three tiers of society are entitled to use the land and resources for their well-being. The strongest entitlements come from matrilineal rights to the land, although the Rongelapese recognize paternal rights to the land as well:

John Anjain: *On Jabwaan, my father is the ri-jerbal. I don't have to ask to use my father's land. On Enewetak, my mother is the alab. All of your mother's family has rights. Even if someone is gone for a long time, his or her rights won't disappear. If other people have rights to the land, they can use it. People who don't have any rights have to ask permission to use it.*[28]

In addition to rights inherited from parents, people obtain rights to the land by permission of the iroij or family, or by marriage:

---

[25] Mike Kabua, peer-review comment, August 7, 1999.
[26] Lijon Eknilang, advisory committee meeting, Majuro, March 2, 1999.
[27] Mitjua Jankwin, interview by Holly M. Barker, Rongelap, March 10, 1999.
[28] John Anjain, interview by Holly M. Barker, Ebeye, March 16, 1999.

Johnsay Riklon: *When you are married, you are entitled to the land of your spouse. It is also understood that relatives can ask for resources.*[29]

Because land is so important to survival, "the Marshallese jealously guard their land rights and will not willingly part with them."[30] The Rongelapese are very clear about the boundaries of their weto property. It is clear, for example, that the extension of weto boundaries incorporate rights to the lagoon:

Emos Jilej: *The boundary of this weto [Tutulonarijet weto], my wife's land, is the small, sharp peninsula you will see further down and the row of pandanus trees over there.*[31]

George Anjain: *My land extends from the middle of this island all the way down to the end. My land covers half of the island. Weto rights extend to the water's edge, including the sand and any exposed areas. Beneath the tidal zone out to 5 miles belongs to the local government. From 5 miles to the 200-mile EEZ [exclusive economic zone] belongs to the national government.*[32]

John Anjain: *People have rights up to the high tide mark where the water ends. The navy explained this to us. Sometimes we just go and take food from the water area on someone's property, but sometimes we ask for permission.*[33]

Everyone who has rights to land also has the responsibility to help harvest or maintain shared resources—including cleaning, planning, harvesting, and the division of food resources. People with land rights are also responsible for deciding what plants will be grown on certain islands and what areas will remain in a wild state. Those who do not have rights to the land can ask the landholders for permission to use the land. They benefit from the resources on the land, and landholders benefit by having their land cared for:

Wilfred Kendall: *People ask permission to work on someone else's land, but it is always expected the land will be returned in good shape. You do what is expected of you or people will have the right to keep you from the land.*[34]

---

[29] Johnsay Riklon, interview by Holly M. Barker, Barbara Rose Johnston, and Stuart Kirsch, Majuro, February 28, 1999.
[30] Jack A. Tobin, *Preliminary Anthropologist Report: Bikini Atoll Survey 1967* (Guam: Office of the High Commissioner of the Trust Territory of the Pacific, 1967), 3, http://worf.eh.doe.gov/data/ihp1c/8366_.pdf (accessed October 25, 2007).
[31] Emos Jilej, interview by Holly M. Barker, Rongelap, March 10, 1999.
[32] George Anjain, interview by Holly M. Barker, Rongelap, March 15, 1999.
[33] John Anjain, interview by Holly M. Barker, Ebeye, March 16, 1999.
[34] Wilfred Kendall, advisory committee meeting, Majuro, March 3, 1999.

John Anjain: *If someone uses your land, they are expected to take care of it, and to maintain it. If they don't, or if they cut trees or wreck the house, you can stop them from making copra. You can even stop a family member from using the land because there is always a leader in the family.*[35]

In some cases, especially in the urbanized areas of Majuro and Ebeye, housing is rented. Renters have the right to occupy the house but do not have the right to use resources associated with the land. They cannot harvest fruits, cut or clear trees, or use other significant resources without express permission from the landowner:

Mike Kabua: *They are merely leasing the house—not the land. The property owner still has the right to collect breadfruit.*[36]

In addition to maintaining the land for their own use, the Rongelapese have a clear perception of the importance of protecting land for future generations. The Rongelapese understand that they are fortunate to have received the gift of property from their ancestors. They in turn look forward to bequeathing it to future generations:

Lijon Eknilang: *It's better to use the word jolet [inheritors] instead of owner. We are the inheritors, but the owner is God.*[37]

The people of Rongelap make a conscious effort to teach thier children and grandchildren where their land rights are so they can continue to claim and use their property in the future. Knowledge of the land, its resources, the family history of use, and weto boundaries is essential to proving a family claim:

James Matayoshi: *My mother shows me where our land is so I can take care of it.*[38]

Emos Jilej: *My grandchildren have never been to Rongelap. I tell them where their land rights are so they will know.*[39]

Mike Jenkins: *I was raised in the United States. My mother was Marshallese. She was an alab of Kwajalein, and my father was American. My mother taught me all about Marshallese land rights. Before I ever came to the Marshall Islands, my mother taught me about the iroij*

---

[35] John Anjain, interview by Holly M. Barker, Ebeye, March 16, 1999.
[36] Mike Kabua, advisory committee meeting, Majuro, March 3, 1999.
[37] Lijon Aknilang, interview by Holly M. Barker, Ebeye, March 16, 1999.
[38] James Matayoshi, interview by Holly M. Barker, Majuro, March 1, 1999.
[39] Emos Jilej, interview by Holly M. Barker, Ebeye, March 10, 1999.

system, land claims, and family rights. My mother chose me out of all ten children to come back to the Marshall Islands to claim my family's land rights.[40]

Cultural Land and Seascapes

Marshallese place-names explain the history of terrestrial and marine property. As illustrated in table 1, atoll, island, weto, seamount, and reef names all remind the Rongelapese of the history and the social and environmental significance of their property.[41]

Weto place-names typically describe physical characteristics or explain why and how people inherited and use the land:

Wilfred Kendall: *Land speaks for everything out here in the island. . . . The name [of the land] speaks to the history, whose rights, and why those rights exist.*[42]

The interviewees know where all the important natural resources are located on the atolls:

John Anjain: *There are places where we lived and places where we would go to stay and gather food. People would stay for years on Buokku, Eniaitok, Jabwaan, Rongelap, Aerik, Mellu, and Yugui.*[43]

One interviewee, Mwenadrik Kebenli, recited a chant (*roro*) about Kapijinamu that teaches listeners to follow waves to the atoll. Waves bouncing off the reef affect wave patterns far from land and help sailors find the reef. The fact that reefs on Rongelap have place-names indicates that reefs were an integral and valued part of the lives of the Rongelapese people. The Rongelapese named reefs and seamounts in the same way they named atolls, islands, and wetos.

One reason northern atolls enjoy such an excellent reputation for fisheries is their proximity to seamounts. Seamounts are large submarine volcanic mountains, rising at least 3,300 feet (1,000 meters) above the surrounding deep-sea floor. Preliminary oceanographic studies have shown that seamounts are biologically rich areas supporting a distinct benthic (bottom-dwelling) community of animals, many of which are unique and do not occur elsewhere on earth. In addition to

---

[40] Mike Jenkins, interview by Holly M. Barker, Gugeegue, March 15, 1999.
[41] See Byron Bender, "A Linguistic Analysis of Place Names of the Marshall Islands" (PhD dissertation, Indiana University, 1963). On file at the Alele Museum, Majuro, and the Pacific Collection, University of Hawaii. See also T. Abo, B. Bender, A. Cappelle, T. DeBrum, *Marshallese–English Dictionary* (Honolulu: University of Hawaii Press, 1976), 499–589.
[42] Wilfred Kendall, advisory committee meeting, Majuro, March 3, 1999.
[43] John Anjain, interview by Holly M. Barker, Ebeye, March 16, 1999.

supporting diverse marine life, seamounts attract pelagic species—schools of large fish, especially tuna, that visit to feed and spawn.

Seamounts are more than a magnet for sea life. They also attract fishermen. Traditional fishing in these areas sustainably supported the community. However, technologically intensive exploitation seriously damages the coral forests and associated marine life found in seamounts worldwide. Seamounts are increasingly being devastated by commercial fishing, especially by trawlers. The United Nations Environment Program and a number of nations have recognized this threat to a unique marine habitat and are attempting to protect seamount resources by designating them as marine reserves.[44]

Of the forty-three named seamounts in the Marshall Islands, some thirty are located in the northern atolls. Seamounts immediately north of Ailinginae are Ruwituntun and Look. Immediately south of Ailinginae (between Ailinginae and Wotho) is Lewonjpoui. Northeast of Rongerik is Lawun-Pikaar. Immediately north of Bikini is Wodejebato. According to Alfred Capelle, some names associated with seamounts were names of sea demons and legendary spirits, both good and evil. They inflicted either good or bad luck on those who knowingly went near them, depending on the intent of the trespasser.[45] Rongelap informants both confirmed the importance of these areas as fishing grounds and expressed their fears that foreign commercial trawlers were destroying these areas in their long absence from Rongelap.[46]

For the Rongelapese and other Marshallese people, the cultural landscape is fixed *and* fluid—both terrestrial and marine. Seamounts, reefs, islands, and lagoons are all significant features of the land and seascape and are valued for many reasons. Marshallese stories of the beginnings of life on earth involve features of the seascape. In many cases, place-names represent the physical manifestations of

---

[44] On May 16, 1999, the Tasmanian Seamounts Marine Reserve was declared under the Australian National Parks and Wildlife Conservation Act 1975. See "Tasmanian Seamounts Marine Reserve," *Australian Government: Department of Environment, Water, Heritage and the Arts,* July 10, 2007, http://www.ea.gov.au/coasts/mpa/seamounts/index.html.

[45] Alfred Capelle, personal communication with author, September 2001.

[46] The United States Geological Survey (USGS) began exploring Marshall Islands seamounts during Operation Crossroads. Much of the research in the 1950s examined the impact of chemical deposition from nuclear weapons tests on seamount reefs. Surveys done in the 1950s and 1960s noted locations of seamounts. Beginning in the 1970s, attention was placed on geological structure, especially the presence of economically valuable mineral resources. In 1999 USGS published a bathymetric map to accompany mineral reports demonstrating the presence of potentially valuable deposits of minerals (James Hein, Florence Wong, and Dan Mosier, *Bathymetry of the Republic of the Marshall Islands* [Menlo Park, CA: USGS, 1999]). The USGS map is most notable for including the Marshallese names of all seamounts, indicating that the Marshallese had extensive knowledge and use of these areas for sociocultural and economic reasons. Marshallese names were provided to Jim Hein by Alfred Capelle, who obtained this information through interviews with Marshallese fishermen. There are forty-three Marshallese names for seamounts on the bathymetric map.

social relationships over time. Land is inherited through family name and valued for its connections to future generations. The name of the land in itself can depict important events and relationships:

> **Mike Kabua**: *The names of some land reflect the possession history of the property: "kited" land is land that an iroij or alab gives to his wife. That land will be passed down through the family of the wife upon her death. . . . Of all the types of land, land acquired by battle is particularly important.*[47]

Morjinkōt (land acquired in battle) literally means "taken at the point of a spear." An iroij gives the land to a warrior to show appreciation for the warrior's battle skills. *Morjinkōt* "was always given by the *iroij* alone to commoners." Rights to the land extended to the family of the warrior; "maternal relatives and paternal relatives both used the land. Maternal relatives have a usufruct right to the land. Paternal relatives could utilize the resources of the land but did not have usufruct rights in the land."[48]

Table 1. Examples of Place-Names on Rongelap, Ailinginae, and Rongerik

Atoll Names
Rongelap: *ron* ("hole," referring to the lagoon) + *lap* ("large") = large lagoon
Rongerik: *ron* ("hole") + *rik* ("small") = small lagoon
Ailinginae: *aelon* ("atoll") + *in* ("in") + *ae* ("current") = atoll in the current

Island Names
Enetaitok: *ene* ("island") + *aitok* ("long") = long island
Enebarbar: *ene* ("island") + *barbar* ("rocky," or "lots of reef") = island with lots of reef
Aeroken: *ae* ("current") + *rok* ("southern") + *en* ("away from speaker") = southern current

Weto Names
Marren: *mar* ("bushes") + *en* ("away from speaker") = land with bushes away from the village
Monbako: *mon* ("house") + *bako* ("shark") = house of the sharks
Aibwej: *aibwej* ("water") = water

Reef Names
The name Kapijinamu, the huge reef in the south of Rongelap, means "the place to fish and get food from the reef." Other reef names are Kijukan, Patelona, Kejenen, Tuilok en Kijenen, and Metalaen.

Source: Land Value Advisory Committee, March 1999.

---

[47] Mike Kabua, advisory committee meeting, Majuro, March 2, 1999.
[48] Tobin, *Land Tenure Patterns,* 29, 29, 31.

All of Rongerik Atoll is *morjinkōt* land. Rongelap's former magistrate, John Anjain, recorded a history of Rongerik Atoll in his magistrate's records.[49] Rongerik was given to Lejkonikik Antoren, an ancestor of Rongerik's current iroij, Anjua Loeak, for heroics in a battle between Kili and Jaluit. For this reason, the property on Rongerik is accorded great respect and great value in Marshallese culture.

Spiritual Values of Land and Seascape

It is exceedingly difficult to use Western terms to convey the spiritual value of land to the Marshallese. All of Marshallese life, culture, and sense of identity emanate from the people's connection to their land. In an April 1999 interview, one Rongelapese youth equated the loss of one's land to hanging oneself with a rope. Comparing the violence levied against the Rongelapese people and their environment to the violent suicides of many Marshallese youth demonstrates that the Rongelapese feel they have been violently removed from the land. In this regard, removal from land takes away purpose and meaning in life and is akin to a violent death for the Rongelapese.[50]

Because of the central importance of land and property to the Marshallese, people have a strong attachment to family land. It gives them a sense of their place in society. Marshallese words that capture the importance of this connection, such as *lamoren,* or "old lineage land," and *jolet,* or "inheritance," translate poorly into English. Connection to the land is expressed throughout the Marshall Islands in national and local government seals, in the constitution, and in the national anthem, popular songs, and stories. The first national anthem of the republic describes the spiritual value of land for the Marshallese:

Ij iokwe lok aelon eo ao
Ijo iar lotak ie.
Melan ko ie im iaieo ko ie.
Ijamin ilok jen e
Bwe ijo jiku emool.
In ao lamoren in deo emman
Lok ne inaaj mij ie.

---

[49] See map 4: map of Rongerik and historical notes by John Anjain.

[50] Suicide rates are unusually high among Marshallese youth. In Australia, aboriginal communities also face high morbidity and mortality rates (much higher than rates for nonindigenous Australians). Medical anthropologist Judith Fitzpatrick (peer-review comment on draft report, October 1, 1999) reports a range of stresses associated with chronic ill health and with high death rates. Constantly holding funerals and the associated rituals drains emotional and financial resources. High death rates seriously inhibit the community's ability to perform social roles and reproduce cultural knowledge. All this contributes to community ill health.

[I love my island
Here where I was born.
The beautiful surroundings and joining together with friends.
I don't want to leave here
This is my true place.
It's my inheritance forever
It's best for me to die here.][51]

The sanctity of certain land is reflected in the range of taboos: walking is forbidden in certain areas, such as medicinal areas (*taban*). Certain areas are set aside for magic; other areas for birthing. On some islands, screaming is forbidden. In some cases, it is taboo to say the real name of a place; nicknames are used out of respect for the place or the events that created the name.[52]

Historically, this attachment to the land appears rooted in the indigenous religion of the Marshall Islands. Although the arrival of missionaries in the nineteenth century effectively replaced the indigenous religion with Christianity, there is evidence of a Marshallese religion devoted to the natural gods, such as Jebro, the god of breadfruit, and Irooj Rilik, the god of fish. People chanted to the sharks before fishing and established sacred bird sanctuaries. In 1958 anthropologist Jack Tobin noted bird sanctuaries in the Marshall Islands existing from "time immemorial" as reserves for birds and turtles.[53] According to Tobin, elaborate rituals dedicated to Lawi Jemo, the *kanal* tree god, accompanied annual food-gathering trips to bird sanctuaries. Chiefs would lead a fleet of canoes from a neighboring atoll to a sanctuary. Upon arrival, women hid under mats in the canoes for fear of bringing bad luck to the gathering of birds and eggs. For the men, "it was tabu to use ordinary Marshallese, [and] the *laroij* (esoteric) language was mandatory."[54] Using special chants, men requested the strength to haul the canoes up on the beach. Once the expedition arrived at the sanctuary island:

> *The chief was the first person to step ashore. Everyone assembled on the beach before proceeding inland and cut a leaf of coconut frond. With the chief leading the way toward Lawi Jemo (the kanal tree), they walked in single file, each individual carefully stepping in the footprints of the person in front of him so that only one set of footprints would appear,*

---

[51] Translated by Holly Barker, with assistance from the advisory committee.
[52] One Marshallese informant reports that *taboo* is not a Marshallese term and may be too strong of a word. In general, rules are known and followed, and specific terms are used for specific sets of rules. Adherence to these rules is not enforced by a strict code of sanctions, as has been described in some other Pacific Islands cultures. The term *taboo* is used here to mean rules that restrict behavior. Members of the advisory committee, when speaking in English, used the term.
[53] Tobin, *Land Tenure Patterns, 50.*
[54] Ibid., 50.

as if only one person had been there. Strict silence was observed on the way to worship Lawi Jemo. *When the group reached the tree, each man placed his coconut leaf over a branch of the tree and then sat down in front of the tree and waited for a breeze to come and blow the leaf off. When this occurred, the* kebbwi in bwil *(ritual name for the chief on this occasion) would announce:* Wurin *(we are lucky).*[55]

The spiritual significance of place was confirmed in our ethnographic interviews; some of the strongest expressions of Marshallese sentiment were voiced in relation to the meaning of atolls and land rights. For the Rongelapese, land represents much more than Western notions of property ownership. Land is the physical framework for a spiritual realm that brings meaning, cohesion, and happiness to people's lives:

Mwenadrik Kebenli: *Without the land, all shatters. Land binds us. I was really happy on Rongerik because it's my place, I grew up there.*[56]

Isao Eknilang: *The three atolls, truthfully are like one person. We looked after one another and fed one another when we were there. . . . The three atolls are like one person in our understanding, and we have to take care of each other.*[57]

Boney Boaz: *On Rongelap, I went everywhere. No one told me I was forbidden from going anywhere. In Ebeye, I just go from my home to work and back again. What is the importance of land? It's so important! The land is what is important. Here are some examples: I planted so many coconut and pandanus seedlings. It used to be great. I didn't watch my father work. I planted the seedlings myself. But now I'm not there to see the trees I planted.*[58]

Dorothy Amos: *Last week [during our quick visit] on Rongelap, I was sad, extremely sad. The atoll is so big. And it's the surroundings where I grew up. I might or I might not live. If they go back there, I will make a trip again. I stood in the church and remembered our good times there. The breadfruit there is all gone. I walked to Jabwaan so I could really see the land. I saw the tombstones and really stared at them. I stared at them for an hour. . . . It was sad.*[59]

Kajim Abija: *We were sad about leaving Rongelap. It's our place. I was accustomed to it.*[60]

---

[55] Ibid., 51–52.
[56] Mwenadrik Kebenli, interview by Holly M. Barker, Ebeye, March 16, 1999.
[57] Isao Eknilang, interview by Holly M. Barker, CMI Nuclear Institute, Majuro, May 7, 1999.
[58] Boney Boaz, interview by Holly M. Barker, Ebeye, March 17, 1999.
[59] Dorothy Amos, interview by Holly M. Barker, Ebeye, March 18, 1999.
[60] Kajim Abija, interview by Holly M. Barker, Ebeye, March 16, 1999.

Isao Eknilang: *Rongelap's local government seal has a picture of the flower for Rongelap and it's arrowroot, and it's not because there was no reason to pick it as Rongelap's flower. The reason is because we ate so much of that food that it was the most important for us.*[61]

In discussing the value of land and traditional means to sustain a livelihood, informants clearly valued their way of life—a life to be proud of, one that allows families and communities to be happily self-sufficient. Money and material are not as important as the ability to live without outside influences:

Mitjua Jankwin: *This is the house where I lived [points to the fallen-down remains of her home on Rongelap, which is nothing more than a concrete floor and rotted planks]. My room was here [points to the front]. My father died in this room [points to another area]. We buried him over there [points to the cemetery down the road].*[62]

Lijon Eknilang: *It makes me sad to come here [the lagoon in front of her grandmother's house on Rongelap], because I used to come here [swimming] all the time with my grandmother. She used to swim everyday in this small pool. . . . I lived in three different places while I was here [on Rongelap]. I moved back and forth between my mother and father's places. Life was so good here. We really had fun.*[63]

Kajim Abija: *Papa said that Rongelap was a place to get younger. You don't get older but you get younger on Rongelap. Rongelap women were known for looking young when they are older, even though they may not be as good looking as women from other atolls when they are younger [laughs].*[64]

Environmental Knowledge and Sustainable Resource Use

George Anjain: *Using resources from all islands and all available atolls is essential to survival. People depend on the ability to use their whole system of islands. In Rongelap, we need to use the northern islands. Survival depends on being able to use everything around us and on sharing food. People got together to sail to distant islands and also to harvest food. Food was then collected and distributed to everyone.*[65]

The people of Rongelap, Rongerik, and Ailinginae, like other Marshallese, had a keen understanding of local environmental conditions, resources, and ecosystemic

---

[61] Isao Eknilang, interview by Holly M. Barker, CMI Nuclear Institute, Majuro, May 7, 1999.

[62] Mitjua Jankwin, interview by Holly M. Barker, Rongelap, March 10, 1999.

[63] Lijon Eknilang, interview by Holly M. Barker, Rongelap, March 10, 1999.

[64] Kajim Abija, interview by Holly M. Barker, Ebeye, March 16, 1999.

[65] George Anjain, advisory committee meeting, Majuro, March 3, 1999.

dynamics. Ecosystem knowledge was essential to survival. Atoll resources provided water, food, building materials, tools, transportation, medicine, toys, and ceremonial items.

The geophysical characteristics of atolls—small spits of land, sparse vegetation, porous substrata that limited the buildup of freshwater aquifers—meant that terrestrial resources were scattered across the atoll rather than concentrated on a specific island mass. The sustainable use of scarce resources required the ability to travel, sparse settlement patterns that encouraged mobility rather than permanent residency, resource-management systems that allowed each household access and use rights to ecologically diverse settings throughout the atoll system, and social and cultural systems that allowed efficient use and equitable distribution of food and other critical resources.

The island ecosystem presented challenging biophysical constraints: scarce land, high vegetative stress from salt-laden winds, constant threat of erosion from wave action and storm surges, relatively infertile soils, minimal freshwater sources, and a tendency toward saltwater intrusion in the few subsurface aquifers present on coral-based atolls. Sustainable use of scarce resources also required the development of resource-management strategies that enhanced ecosystemic viability while encouraging greater productivity. People enriched soil fertility with fish remains. They carefully nurtured pandanus, breadfruit, and coconut trees. They used woven mats and baskets to shelter seedlings from the wind and fierce sun. They regularly pruned mature trees, carefully piled and burned the fronds and other debris, and reworked ashes into the soil. They used rocks and coral rubble to build small retaining walls around the root systems of trees growing near the shore. Rubble retaining walls captured plant debris, allowed the buildup of soil around root systems, and served as protective barriers during periodic tidal surges accompanying tropical storms. These traditional agroforestry practices are still used today and illustrate the essential role played by humans in the maintenance of the Marshall Islands ecosystem.

Some Rongelapese believe that people and plants are symbiotically linked. After observing that the trees and plants on Rongelap have no flowers or fruit, Lijon Eknilang explained, "Plants don't grow without people. We make the plants happy and they make us happy."[66]

Without exception, and despite long absences from their lands, all interviewees were able to clearly identify and describe the resources and use strategies critical to survival on Rongelap, Rongerik, and Ailinginae.

In reviewing traditional patterns of resource access and use, interviewees expressed keen frustration in their current ability to teach the younger generation

---

[66] Lijon Eknilang, interview by Holly M. Barker, Rongelap, March 9, 1999.

the information and skills needed for survival on an isolated atoll. Knowledge that informants identified as being important to pass on to the next generation included information on survival and self-sufficiency, sustainability, storing and preserving resources, sharing of resources, medicine, legends or *bwebwenatos,* navigation, and strategies for coping with famine, drought, and other dangers.

Drinking water sources include springs and ponds fed by subsurface aquifers, rainwater caught and stored in giant clamshells (*emok*), holes in the ground from fallen coconut trees, and rainwater caught from tin roofs or in catchments first paved by the Japanese. Despite long absences, and in the case of Rongerik despite lack of personal experience, interviewees were able to describe the location of potable water sources on Rongelap, Rongerik, and Ailinginae atolls:

John Anjain: *On Jabwaan, we drank the well water. It was less salty than the well water on Eniaitok or Rongelap. Rongerik had good ground water. Ailinginae had one well. Mwenlap weto on Jabwaan has the best water. It is a water source from long ago.*[67]

Lijon Eknilang: *We had different ways to quench thirst . . . coconut milk, pandanus, we used the big roots from the pandanus tree to get at water.*[68]

Wilfred Kendall: *Anyone can get water, but you must ask permission. Rainwater is more reserved for family.*[69]

Fish is central to the Marshallese diet. The Marshallese–English dictionary has sixty-six entries depicting a wide range of fishing methods used for different weather, times of the day, and species of fish. People fished with long lines (*mueo*), nets (*ok*), and traps (baskets of different sizes). Lobsters were snared. Rainbow runners and dolphins were hunted with lines, lassoed, and beached. The Rongelapese consider themselves masters at fishing and pride themselves in their knowledge about where and how to fish. Rongelapese men sailed throughout their three-atoll system to exploit different fishing areas. On Rongerik, one of three ocean beaches was set aside for women to fish, usually with nets.[70] The entire community helped chase migratory schools of fish onto reefs and into shallow storage ponds. The Rongelapese used fish as food and to pay tribute to the *alab* and iroij. They also used fish bones and scraps to fertilize their land.

John Anjain: *We got food from all of the islands on Rongelap. There was lots of fish and lobster on Buroku Island. It was a place to get our meats.*[71]

---

[67] John Anjain, interview by Holly M. Barker, Ebeye, March 16, 1999.
[68] Lijon Eknilang, advisory committee meeting, Majuro, March 2, 1999.
[69] Wilfred Kendall, advisory committee meeting, Majuro, March 2, 1999.
[70] Lijon Eknilang, advisory committee meeting, March 3, 1999.
[71] John Anjain, interview by Holly M. Barker, Ebeye, March 16, 1999.

Boney Boaz: *On Rongelap, Karuwe, and Boarok islands in the north were the best places to fish.*[72]

Many of the interviewees discussed the importance of birds and bird eggs to their diet. In fact, the people on Ailinginae were busy collecting birds and bird eggs when they experienced the fallout from the Bravo test. Bokankaer, Enebarbar, Enealo, and Eniaitok islands were particularly known for having many birds, including *ak* (frigate birds), *kalo* (brown boobies), *kear* (terns), and *pejwak* (brown noddies). The interviewees were able to identify favorite locations for nesting birds and describe techniques for capturing them:

Lijon Eknilang: *We would break the wings of birds when they were young, so they wouldn't fly away [enabling people to go back and get them when they were older]. Not all the young have their wings broken. We also collected the eggs of birds to eat. Birds were found on certain islands. Some types of birds were best to catch at nighttime.*[73]

On Rongerik, people used to collect the eggs from many birds, including the *oo, ak, kalo, kear,* and *pejwak.* Jipedbao Island (whose name "the place with birds") on Rongerik was known for its birds.

Mitjua Jankwin: *On Ailinginae there were plenty of oo birds to eat. We used to fill buckets with oo.*[74]

All five of the world's sea turtle species nest in the Marshall Islands. Turtles were found all over Rongelap, Rongerik, and Ailinginae. There are no seasons for turtles. Again, the interviewees knew the best locations for turtles. Bock Island on Rongerik is particularly well-known for turtles.

George Anjain: *Turtles used to lay eggs everywhere on the atolls, even on populated islands.*[75]

Nerja Joseph: *Rongerik is known as the atoll of birds and turtles.*[76]

Mitjua Jankwin: *Turtles used to lay their eggs on Rongelap. There were turtles on all the side islands.*[77]

---

[72] Boney Boaz, interview by Holly M. Barker, Ebeye, March 17, 1999.
[73] Lijon Eknilang, advisory committee meeting, Majuro, March 2, 1999.
[74] Mitjua Jankwin, interview by Holly M. Barker, Rongelap, March 10, 1999.
[75] George Anjain, advisory committee meeting, Majuro, March 3, 1999.
[76] Nerja Joseph, interview by Holly M. Barker, Majuro, March 8, 1999.
[77] Mitjua Jankwin, interview by Holly M. Barker, Rongelap, March 10, 1999.

Emos Jilej: *Rongelap was always known for its bounty of food. It was number two in turtle after Kwajalein. There were turtles on all the small islands.*[78]

Giant clams and other clam species found in the reef were important sources of food. Clamshells were used soften the fibers woven into mats. The shells of giant clams were used to collect and store rainwater. Women and children often gathered smaller clams found along the shoreline. Men would collect the larger clams found in deeper waters.

Lijon Eknilang: *On Rongelap, there were always lots of clams on the reef where my grandmother lived, near the church.*[79]

John Anjain: *Ailinginae was known for its clams and birds. We would ask for permission to make trips to gather them.*[80]

Coconut crabs are considered a delicacy by Marshallese. Islands such as Arbar on Rongelap Atoll had crabs everywhere, and the Rongelapese ate coconut crabs on a consistent basis. They also ate smaller crabs that lived in the rocks and coconut husks by the shore. Other foods included lobsters, octopuses, and pumpkins.

John Anjain: *The reefs on all three atolls were full of lobsters and octopus. Anyone could go to the reefs and collect them because the reef food belonged to everyone.*[81]

Lijon Eknilang: *There was pumpkin all the time. It has no season.*[82]

Kajim Abija: *When people traveled from Ailinginae to Rongelap at Christmastime, they would always bring pumpkin.*[83]

Coconuts provided not only food but also materials important to survival. As the Marshallese advisory committee noted, there are one thousand different uses for the coconut tree. The Marshallese depended on the coconut tree for food, drink, building materials, toys, and money earned from selling dried coconut meat. The Rongelapese traded dried coconut meat, or copra, for cash or for supplies from the ships that stopped in Rongelap every three to six months.

---

[78] Emos Jilej, interview by Holly M. Barker, Rongelap, March 10, 1999.
[79] Lijon Eknilang, interview by Holly M. Barker, Rongelap, March 10, 1999.
[80] John Anjain, interview by Holly M. Barker, Ebeye, March 16, 1999.
[81] John Anjain, interview by Holly M. Barker, Ebeye, March 16, 1999.
[82] Lijon Eknilang, interview by Holly M. Barker, Kwajalein Island, March 19, 1999.
[83] Kajim Abija, interview by Holly M. Barker, Ebeye, March 16, 1999.

Johnsay Riklon: *While the United States might look at a coconut tree and see the value of the copra, we see medicine, toys for our kids, food, weaving materials, sails, and canoes. Nothing is wasted. One coconut tree, similarly to a pandanus or breadfruit tree, can almost support a family with all of its needs.*[84]

Pandanus, breadfruit, and arrowroot were other staple food sources. Breadfruit can be prepared in a variety of ways. Arrowroot was so abundant that the Rongelapese used it to make flour. Pandanus and breadfruit are seasonal foods; arrowroot is not. The Rongelapese distributed these foods as gifts to their relatives and their iroij, who in turn shared them with people throughout the Marshall Islands. The Rongelapese often gave gifts of staple foods and later received gifts of other cultivated foods or supplies. When the supply ships sailed to Rongelap, Rongerik, and Ailinginae, people bought flour, sugar, rice, and other foods, but the preservation techniques used to prepare pandanus, breadfruit, and arrowroot ensured that the people did not go hungry if the supply ships did not arrive.

Pandanus was important for making certain types of food for the iroij, such as *jaankun*. People ate the fruit raw or boiled it and used the softened meat for baby food, baked goods, and candy. People also wove the leaves of pandanus trees into sleeping mats. With water scarce, pandanus, like coconut, quenched thirst. During times of extreme drought, people would steam pandanus roots in a fire until the water came out.

Nonedible plants were used for windbreaks, tools, spices, and particularly medicine. Some knowledge of medicine was passed down in a family, and the knowledge to make certain medicines was possessed by only a few people. Some parts of a plant were more powerful than others. Some plants were deadly at certain times of the year. Some means of ingestion could cure, while others could harm or even kill.

Knowledge about medicine was considered powerful. People used virtually every indigenous species of plant for medicinal purposes:

Lijon Eknilang: *Everything that grows is medicinal. Even grass is used for kids, and as reproductive medicines. But Marshallese medicines are never to be used on menstruating women. When the people lived on Rongelap, Rongelap used to be a place for making medicine.*[85]

The interviewees felt that Rongelap, Rongerik, and Ailinginae were very good areas for acquiring medicine. Although Marshallese medicine was free, healers benefited from the power and prestige that their knowledge accorded them and from the

---

[84] Johnsay Riklon, interview by Holly M. Barker, Barbara Rose Johnston, and Stuart Kirsch, Majuro, February 28, 1999.
[85] Lijon Eknilang, advisory committee meeting, Majuro, March 3, 1999.

food and other expressions of appreciation given by patients. *Ri-bubu*, usually men, were particularly strong healers.

It is clear from the interviews that legends and storytelling were important means to transmit knowledge about the environment that was essential to survival. For example, local stories told listeners the best locations for catching eels, finding other important resources, and navigating the islands:

Mwenadrik Kebenli: *On the ocean side of Rongerik, when you look out at the ocean, it looks like there are men out there fishing, and they are looking for eels. The rocks are shaped like that.*[86]

John Anjain: *An old man, Jelan, saw an* akwolej *bird digging into the ground. When he went to see what it was, he found water.*[87]

Kajim Abija and Lijon Eknilang: *In order to find Rongelap, sailors would follow the North Star, Limanman. Limanman is included on the Rongelap local government's seal to help people find the atoll.*[88]

John Anjain: *Eniaitok, an island known for its birds, is shaped like a bird. There is a story of a boy, Leok, who threw a rock at the bird, Jebtaka, and knocked it to the ground. The north and south sides of the island are the wings, and the central part of the island, Jibiken, is the tail of the bird. The large stone in the reef on the northern part of the island is the stone used to knock the bird down.*[89]

Knowledge about dangers in the environment saved the Rongelapese from potential hazards or threats to their lives. This knowledge was critical to the well-being of the community. Children were taught about species of fish and plants that were dangerous to consume or touch. The adults on Rongelap also taught children about locations with many sharks:

Shirley Kabinmeto: *As children, we didn't go swimming at the end of the island where I lived, Jabwaan, because Jabwaan is full of sharks. Jabwaan is at the end of the island, near a deep pass where the sharks enter the lagoon from the ocean side.*[90]

Navigational knowledge was transmitted in roro, or ancient chants. They instructed sailors about the hazards of reefs and how to navigate the islands. Locally based knowledge about navigation was crucial to the survival of the Marshallese.

[86] Mwenadrik Kebenli, interview by Holly M. Barker, Ebeye, March 16, 1999.
[87] John Anjain, interview by Holly M. Barker, Ebeye, March 16, 1999.
[88] Kajim Abija and Lijon Eknilang, interview by Holly M. Barker, Ebeye, March 16, 1999.
[89] John Anjain, interview by Holly M. Barker, Ebeye, March 16, 1999.
[90] Shirley Kabinmeto, Interview by Holly M. Barker, Rongelap, March 10, 1999.

People could not live solely from the resources of one island and depended on navigation to maintain their social and political relationships and to gain access to the full range of natural resources necessary for survival. The Marshallese were legendary, even among Pacific Islanders, as being the best navigators.[91]

The Rongelapese relied on their navigational abilities to access the small islands on Rongelap, Rongerik, and Ailinginae. This was particularly true for Ailinginae ("atoll in the current"), which has one of the most difficult currents to navigate. The people who went to Ailinginae were expert sailors and navigators.

Boney Boaz: *I know how to sail an outrigger canoe. I've sailed to every single island on Rongelap, which is about 30 miles across. . . . On Rongelap, we grew up learning sailing and navigation.*[92]

Natural resources provided the essentials for biological well-being such as food, shelter, and water. During the collection of ethnographic data, one interviewee explained the importance of land in terms of self-sufficiency and how the land and the resources provided for people:

Dorothy Amos: *The importance of land? We eat from it. We drink from it. We make handicrafts from it. We can make foods and things to drink from it. We make enra, Marshallese plates for our one-year birthday celebrations. We also make copra that people can sell. We prepare the fronds from the coconut tree for sitting on and making fires. We also make brooms for the houses. And pandanus, the ancestors made clothes from the coconut and pandanus. Women use the pandanus leaves to make sleeping mats, things to lie on. . . . And many kinds of food. There is pandanus time, and breadfruit. Wonderful food. Boats—outriggers. We make Marshallese tin by weaving coconut fronds. Coconut husks to make ropes. They don't cost anything. Today they are really expensive.*[93]

Ethnographic interviews also involved considerable discussion of the critical role played by natural resources in providing the material essentials for social well-being. First-year birthdays (*kemem*), marriages, funerals, and iroij visits were important occasions where community members met as a group, celebrated, and socialized. At one year, a child had a much greater chance of surviving and therefore was recognized as part of the community. Gifts and food were exchanged to celebrate the well-being of the child:

---

[91] Krieger, *Island Peoples of the Western Pacific*.
[92] Boney Boaz, interview by Holly M. Barker, Ebeye, March 17, 1999.
[93] Dorothy Amos, interview by Holly M. Barker, Ebeye, March 18, 1999.

Mike Kabua: *It is important to have materials to weave mats for funerals, as well as important occasions like iroij visits, the first-year birthday party (kemem), weddings, and Christmas. It is also essential to share food with everyone on these occasions, as the gifts and the food make people feel closer together as a community.*[94]

Mike Kabua: *During funerals, people come together and revisit and learn their family connections and place in society. People give gifts to burial areas. Community members place white rocks around grave sites as a symbol of purity. People come together at the eoreak to forgive each other . . . a time to forget hatreds and to forgive any misdeeds of the departed. People know where the burial areas are. They also know where the sea burials took place.*[95]

Lijon Eknilang: *Certain flowers are used for the iroij during visits, such as the* kano *[fern]. People always put the flowers on a necklace so they don't touch the head of the iroij.*[96]

Mike Kabua: *Food is prepared in advance of the visit. Fish comes from the coral heads designated for the iroij. Certain types of foods are given to the iroij, such as turtle, coconut crabs, preserved breadfruit and pandanus, arrowroot, dolphins, and certain species of fish. The only type of coconut the iroij can drink is the* nibarbar. *Food is placed in a large, woven basket, or* kilek, *that only the alab can bring to the iroij. The iroij collects food as he goes from island to island.*[97]

Flexible Patterns of Resource Use—Sustainable Living on Atoll Ecosystems

A major factor that provided the Rongelapese with a flexible, fluid means of gathering the resources necessary for survival was full access to a range of cultivation options from the numerous small islands of the three atolls. The Rongelapese accessed plants, terrestrial animals, birds, and marine and reef life from Rongelap, Rongerik, and Ailinginae. This range of options allowed the Rongelapese to live within their environment by ensuring that they did not deplete available resources in any single area. Multiple resource-gathering options also enabled the Rongelapese to adjust to seasonal and climatic variations.

To protect the resource base, sustainable resource use was an important aspect of food gathering on Rongelap, Rongerik, and Ailinginae. For example:

George Anjain and Lijon Eknilang: *The Rongelapese know to only fish for two days in one spot and then move. With the coconut crab, we didn't used to take the females,*

---

[94] Mike Kabua, advisory committee meeting, Majuro, March 3, 1999.
[95] Mike Kabua, advisory committee meeting, Majuro, March 3, 1999.
[96] Lijon Eknilang, advisory committee meeting, Majuro, March 3, 1999.
[97] Mike Kabua, advisory committee meeting, Majuro, March 3, 1999.

and we didn't take too many from one place. We never took small crabs or pregnant crabs. As for the turtles and birds, we never took all their eggs. We didn't break the wings of all the older birds either, because we wanted to leave some older birds to lay more eggs. These principles are also true for trees and plants, because if you don't clean them and take care of the dead flowers or fruits, the trees and plants get sick and don't grow well.[98]

Sustainable gathering techniques were also used with the oo birds:

Mitjua Jankwin: *We broke their wings when they were young. We ate the younger birds so the adult birds could make more birds. We only ate them during the birthing season.*[99]

Because of the limited seasons of their primary food sources, it was critical for the Rongelapese to be familiar with techniques for preserving and storing food, in particular staple crops such as breadfruit, pandanus, and arrowroot, but also water:

Lijon Eknilang: *Breadfruit appears in the* rok *season, from July through December. Pandanus and arrowroot appear in the* anonaen *season, from December through July. People got together to harvest arrowroot, breadfruit, and other foods to preserve them. We salted, dried, and grated pandanus for preservation. Jekaka is dried pandanus that is grated like a powder or flour so it can last for years.*[100]

Timako Klonij: *Young girls like me learned to make* janwin *[preserved pandanus] and other foods from pandanus.*[101]

Lijon Eknilang: *Rainwater was carefully collected and stored. People placed basins, or emok, at the bottoms of coconut trees to catch runoff rainwater. These basins were usually made of clamshells or carved wood.*[102]

Because food and water are critical to life but limited on a coral atoll, the sharing of resources was essential to survival. The Rongelapese worked together to harvest and prepare food. They shared water and food resources with their families and neighbors. The interviewees discussed how they would come together and work during breadfruit season, from July through December, to make preserved breadfruit for ceremonies and family consumption, and to send to distant family members and iroij. Other resources were shared too:

---

[98] George Anjain and Lijon Eknilang, advisory committee meeting, Majuro, March 3, 1999.
[99] Mitjua Jankwin, interview by Holly M. Barker, Rongelap, March 10, 1999.
[100] Lijon Eknilang, advisory committee meeting, Majuro, March 3, 1999.
[101] Timako Kolnij, interview by Holly M. Barker, Ebeye, March 18, 1999.
[102] Lijon Eknilang, advisory committee meeting, Majuro, March 3, 1999.

Wilfred Kendall: *Because of its central importance, water was shared perhaps more than any other resource. It was understood that if you have a good well, people can come and ask to use it, or just take it.*[103]

Mitjua Jankwin: *On Monluel weto, there was a big fence with chickens in it. The birds belonged to the family, but people would ask for them. When there were many birds, sometimes they would hand out a bird to each family.*[104]

Due to droughts, the seasonality of important resources, and the intense labor required to secure food, people periodically faced times of famine. Because the Rongelapese could depend on the small islands of three atolls and ocean resources, they could gather resources from multiple areas. Multiple access sites and knowledge about their environment allowed the Rongelapese to survive times of famine and drought relatively well:

Kajim Abija: *There was always pandanus and coconuts. During the times of hunger, we were not really hungry because there was so much of the pandanus.*[105]

Boney Boaz: *We used to eat coconut in times of hunger. If I could have eaten those foods, like [restricted coconut] crab, during the famines [after resettlement in 1957], I wouldn't have been so hungry.*[106]

Timako Kolnij: *We relied on sprouted coconut and things from the ocean—large clams, crab, small clams, fish—they were adequate in the times of hunger. We ate mature coconut and drank the drinking coconuts. . . . As a water substitute, people would suck on the coconut husk or the roots of the plant to get the liquid out. We also drank the young coconuts, or ni.*[107]

Despite the many years that have passed since they last lived in Rongelap, informants demonstrated a keen and thorough knowledge of the best times, places, and techniques for catching various species of fish. Particular species run at certain times of the year:

Lijon Eknilang: *In January and February, the mackerel [mwilmwil] run. In November and December the grouper fish [lojepjep] run.*[108]

[103] Wilfred Kendall, advisory committee meeting, Majuro, March 2, 1999.
[104] Mitjua Jankwin, interview by Holly M. Barker, Rongelap, March 10, 1999.
[105] Kajim Abija, interview by Holly M. Barker, Ebeye, March 16, 1999.
[106] Boney Boaz, interview by Holly M. Barker, Ebeye, March 17, 1999.
[107] Timako Kolnij, interview by Holly M. Barker, Ebeye, March 18, 1999.
[108] Lijon Eknilang, interview by Holly M. Barker, Kwajalein Island, March 19, 1999.

When the fish were running, the Rongelapese would work as a group to catch as many fish as possible and to smoke and preserve them so they could be eaten for many months and sent to the iroij and family members in distant places.

Taboos and Resource Management

Customary rules and traditions are an essential component of the Marshallese psyche. Rules that restrict behavior (taboos) are traditionally defined and imposed by internal actors in ways that establish and regulate social, political, and economic relationships. Traditional taboos make sense. The taboo and the rationale for a taboo are often embedded in stories, legends, and sayings. Some taboos support the three-tiered social structure by demonstrating respect for the iroij. For example, it is taboo to touch the head of an iroij, stand or sit above or higher than the iroij, or pass between the iroij's house and the lagoon. Other taboos are for health and safety reasons. For instance, it is taboo to have sexual relations for several months after a woman gives birth, or for women to climb coconut or other trees. Some fish are taboo during certain seasons (and if eaten when taboo, will result in fish poisoning). Still other taboos enable people living on very small parcels of land to cooperate and live together peacefully. It is inappropriate, for instance, to express anger or grievance directly with someone. Instead, the use of a third party to relay information and mitigate the disagreement is encouraged. Some taboos reinforce the roles of women and men in society:

> **Lijon Eknilang**: *It is taboo for women to be on the beach when some fish show up. . . . On Rongerik Island, there are three sides, one of which is reserved for women to swim, fish—a place that is taboo for men.*[109]

Many taboos restrict free access but establish permissive use. Use of critical resources—such as drinking water, wild plants, crabs, and seashells—and fishing off coral heads are generally unrestricted for family members and accessible with permission by others.

> **Mike Kabua**: *I guess that's why we respect each other. Someday you will need something from your neighbor.*[110]

Many taboos regulate resource access and use. A number of islands set aside for the iroij are migratory bird nesting sites:

---

[109] Lijon Eknilang, advisory committee meeting, Majuro, March 3, 1999.
[110] Mike Kabua, advisory committee meeting, Majuro, March 2, 1999.

Wilfred Kendall: *There are areas we know that are set aside specifically, and you don't clear the land. Animals need food from plants. These islands are set aside to leave as they are. The knowledge [of these areas] is handed down through the generations.*[111]

Mike Kabua: Kinbit *is the word for the rules and regulations about when you collect coconut crab. . . . Only the largest males are collected, no females. Nowadays I see people eating female coconut crab because they don't know how to tell. I say, "You know you're eating female coconut crab?" They say, "How do you know?" I turn it over and show them . . . and they say "oohhh!" [laughs].*[112]

Taboos established the rules about what could and could not be used, and why. Some foods and locations were taboo for all but the iroij, some taboo for all but the *alab*. Some foods were taboo for all but the ri-jerbal. Many taboos helped shape seasonality, protecting plants and animals during vulnerable periods and allowing use without overexploitation.

Concluding Discussion

The traditional way of life in Rongelap, Rongerik, and Ailinginae atolls involved social, cultural, and economic activities, values, and meanings that supported a vibrant, marine-based, self-sufficient, and sustainable way of life. Inherited access and use rights, and associated stewardship responsibilities, covered atoll islands, reefs, lagoons, and the myriad of resources within. The natural resources of Rongelap, Rongerik, and Ailinginae provided for the daily and long-term needs of the household and the community. Food, shelter, transportation, tools, and toys were all derived from locally accessible materials. According to informants, and the records kept by Rongelap magistrate John Anjain, in the years immediately preceding the U.S. nuclear weapons testing program, relatively few provisions—kerosene, tin roofing, lamps, cigarettes, matches, thread, needles, sugar, rice, and flour—were purchased, and the lack of these items did not severely impact household health. These were supplementary provisions acquired by the Rongelapese by trading with or selling copra and handicrafts to the ships that passed by their islands. Men processed and sold copra.[113] Women exchanged handicrafts for goods. Rongelap was the first area to establish a woman's group:

---

[111] Wilfred Kendall, advisory committee meeting, Majuro, March 2, 1999.
[112] Mike Kabua, interview by Holly M. Barker, Majuro, July 18, 1999.
[113] Tobin, *Land Tenure Patterns;* Mason, "Ethnology of Micronesia."

Lijon Eknilang: *We had a co-op and used to do all kinds of work, such as making copra, pumpkin, and limes. The name of this group was White Rose, named after something rare and beautiful.*[114]

In addition to providing for the day-to-day economy of the household and community, natural resources provided the Rongelapese with the means to create and sustain social relationships. Resources were exchanged for other resources or for labor, or to mark important cultural occasions. The rights and responsibilities associated with these resources both reflected and reinforced the basic rules of society. As will be further illustrated in later chapters of this report, environmental contamination from nuclear weapons testing not only damaged the health and viability of the natural resources and those who relied upon them but also denied the Rongelapese the means to produce and reproduce their self-sufficient lifestyle. The social, cultural, economic, political, and psychological consequences of this damage and loss are complex and profound:

Lijon Eknilang: *When we move from Rongelap, it's like we are using somebody else's bedroom, and one of these days these people will come and ask for their room back.*[115]

Mike Kabua: *Land speaks for everything out here in the islands. . . . [Without land] you lose your respect for your elders, community leaders. Customs mean respect. Without land, you don't have any fear of punishment. . . . Authority comes from land; respect for the elders comes from land.*[116]

Lijon Eknilang: *People who were born outside of Rongelap have little sense of belonging. Their sense of identity is through their parents, not from having lived [on Rongelap], not from experience.*[117]

In its review of compensation claims for the people of Bikini and Enewetak, the Nuclear Claims Tribunal has awarded compensation for loss of land based on valuation of land derived from a record of leases in Majuro, Kwajalein, and other atolls. Such valuations reflect one aspect of economic value—the value of temporal occupation of dry acreage. They do not reflect or address the damages associated with the loss of the wider realm of arboreal and marine resources that sustain the Marshallese way of life. The data contained in this review demonstrate the need to consider and address:

---

[114] Lijon Eknilang, interview by Holly M. Barker, Rongelap, March 10, 1999.
[115] Lijon Eknilang, advisory committee meeting, Majuro, March 2, 1999.
[116] Mike Kabua, advisory committee meeting, Majuro, March 3, 1999.
[117] Lijon Eknilang, advisory committee meeting, Majuro, March 2, 1999.

- Natural resource damage and loss of lagoons, reef heads, clam beds, reef fisheries, and turtle and bird nesting grounds
- Cultural resource damage and loss of access to family cemeteries, burial sites of iroij, sacred sites and sanctuaries, and *morjinkōt* land.
- The consequential damages to this and future generations produced by the loss of access to and ability to use a healthy ecosystem.
- Consequential damages, including the inability to interact in a healthy land- and seascape in ways that allow the transmission of knowledge and the ability to sustain a healthy way of life.

Part 3

Chain of Events and Critical Issues of Concern

The U.S. nuclear weapons testing program in the Pacific permanently altered the lives and future of the people of Rongelap, Rongerik, and Ailinginae atolls. The Rongelap people experienced personal health injuries and involuntary displacement when fallout from the largest thermonuclear weapon ever tested by the United States (Bravo test) forced their physical evacuation from 1954 to 1957. They lost access to a viable healthy ecosystem and lost their ability and rights to safely live in their environment when they were resettled on Rongelap (1957 to 1985). And they became exiles (1985 to the present) when they were finally provided evidence of hazardous levels of radioactive contamination throughout their homelands. These events and critical issues of concern are identified and briefly contextualized with testimony from Rongelap informants. Evidence that substantiates informant accounts and consequential damages is found in the declassified documents cited in this section and other portions of this report.

Evacuation from Rongelap to Lae in 1946

As a precautionary step, the U.S. Navy evacuated the Rongelapese from their home islands from May to August 1946 for the Able and Baker weapons tests on Bikini Atoll. It was felt at the time that the Rongelapese were so close to the testing area that it was best to remove them as a precaution. In May 1946, 108 Rongelapese and the people from Wotho were temporarily taken to Lae on LST 1108.[1] The U.S.

---

[1] John Anjain, record book, 1946, 118. Magistrates were elected for every atoll, and they acted as the contact people for the Trust Territory of the Pacific Islands government. Anjain was magistrate of Rongelap during the U.S. weapons tests. As magistrate, he kept record books detailing important

government contemplated the permanent resettling of downwind populations to Lae, but the Rongelapese were returned to Rongelap shortly after their stay on Lae.[2]

Almira Matayoshi: *We were sad when we left Rongelap, because we had no idea what we were going to face when we got to Lae. We ran out of food on Lae. There was rice, but nothing to go with it. We relied on the people of Lae to give us what little they could with so many people on their island . . .*

When we returned from Lae to Rongelap, we had to start from the beginning in terms of food production, and getting our households back in order. For example, we missed arrowroot season, so we didn't have time to prepare the foods that would store for long periods of time and get us through the more difficult times. We came back between the food seasons, so we had very little to eat. There were also more of us in our group because we had children on Lae, and some of us married. Our houses needed to be redone. Some of the roofs were gone. The navy didn't stay and help us.[3]

Isao Eknilang: *I clearly remember the time. I knew there was testing of some sort going on, but I didn't really understand it. It was difficult because we lived in tents for several months. We had no privacy, and many aspects of Marshallese custom were broken, particularly those relating to adults changing their clothes or trying to leave to go to the bathroom in front of children of the opposite sex . . .*

Another custom that people began to violate for the first time I remember was not sharing food with people. The navy gave some food to us Rongelapese, but it was not adequate. I remember a fistfight that broke out over whether or not to share our food resources. Some people thought we had to share our food with the people of Lae and Wotho, and others thought we shouldn't because the resources were inadequate for everyone. I remember being stunned that adults would fight over this. In addition to food, there was not enough water for everyone, and the navy didn't bring in water supplies when we asked.

----

activities, events, and decisions within the community. Anjain produced five record books (one taken by a U.S. scientist and never returned, Anjain complains). In April 1999, Holly Barker received permission to photocopy relevant portions of one of the books now held by George Anjain, including the map of Rongerik (map 4).

[2] Gordon M. Dunning, ed., *Radioactive Contamination of Certain Areas in the Pacific from Nuclear Tests—A Summary of Data from Radiological Surveys and Medical Examinations* (Washington, DC: Atomic Energy Commission, January 1957), chapter 6, page 2590.2, http://worf.eh.doe.gov/data/ihp1d/1554_f.pdf (accessed December 17, 2007). DOE and DOD documents were classified and were not available to the Marshallese government or the general public during negotiations for the Compact of Free Association between the Republic of the Marshall Islands and the United States. The majority of these documents were declassified between 1994 and 1999 as part of the Clinton administration inquiry into the use of human subjects in radiation experiments.

[3] Almira Matayoshi, interview by Holly M. Barker, Honolulu, June 13, 2001.

We tried to fish on Lae to supplement our diets, but the fish were much smaller, and it was hard to catch enough fish to feed so many people. On Rongelap, people had three entire atolls to cultivate food. On Lae, an island a fraction the size of Rongelap, there were not enough resources to feed three atoll populations [Rongelap, Wotho, and Lae], despite the food the Navy did supply . . .

While on Lae, we worried about our homes, and what would happen to our animals and plants when we weren't there to tend to them. We children attended school while on Lae, but the school was extremely overcrowded with the addition of the children from Rongelap and Wotho. Also, we didn't earn money while on Lae, because we didn't own the coconut trees, so we couldn't produce copra. By the same token, we lost the opportunity to make copra on our own land during that time . . .

When we returned to Rongelap from Lae . . . it was clear that the U.S. had used our land and houses. Some of our sailing canoes were found on distant islands, and others disappeared and were never returned. We even had a large sailing canoe that could hold twenty people. The canoe was gone when we returned. . . . The canoe is very essential to gather food, housing materials, to produce copra and it takes at least six months to build a walab.

Also, the turkeys that we used to care for and keep on Rongerik were all gone when we returned. Similarly, the pumpkin that was on Ailinginae was gone. Also, our houses were a mess. The thatch roofs were falling apart, and water leaked into many of the houses. . . . The bushes were all overgrown, and we had no local food ready to use. We were not able to get local food for some time. The U.S. Navy left two U.S. boats filled with packaged food. We ate the food, but later on we were concerned because we heard that the boats had been at Bikini for the tests and were probably contaminated.[4]

SUMMARY OF CONSEQUENTIAL DAMAGES

Conditions of life during the involuntary resettlement on Lae were harsh, with food and water in short supply. Traditional customs were violated as people were forced to compete for increasingly scarce resources. The people of Rongelap experienced fears and anxiety about their exile, and concern for their homes. Upon return to Rongelap, the people confronted extensive damage to homes and property, inadequate food supplies, and fears that provisions left on U.S. Navy boats used to support test operations at Bikini were contaminated. Furthermore, they had lost the opportunity to generate an income by cultivating their resources.

---

[4] Isao Eknilang, interview by Holly M. Barker, Honolulu, June 13, 2001. This testimony was reviewed by the informant, and minor details were corrected. Translation and details submitted by Senator Abacca Anjain-Madison, September 14, 2001.

Damage and Continued Loss of Access to Rongerik

Rongerik Atoll is *morjinkōt* land, given by the iroij alone to commoners to show appreciation for a warrior's battle skills.[5] Rights to the land extend to the family of the warrior, as "maternal relatives and paternal relatives both used the land. Maternal relatives have a usufruct right to the land. Paternal relatives could utilize the resources of the land but did not have usufruct rights in the land."[6] Rongerik was given to Lejkonikik Antoren, an ancestor of Rongerik's current iroij, Anjua Loeak, for heroics in a battle between Kili and Jaluit. For this reason, the property on Rongerik is accorded great respect and great value in the Marshallese culture (see map 4).

Population movement between atolls was restricted by the Japanese military during World War II. In 1943 the Japanese moved all people residing on Ailinginae and Rongerik to Rongelap. Rongelap's magistrate John Anjain recorded this event, noting the names of the nine people moved off of Rongerik on January 12, 1943, and the twelve people moved off Rongerik on July 3, 1943.[7] When the United States liberated the Rongelapese from Japanese control on April 4, 1944, the Rongelapese declared this date their Liberation Day.[8] The U.S. Navy continued the ban on access to and use of Rongerik Atoll. It used Rongerik for monitoring weather and tracking fallout.

In 1946 the U.S. government relocated 181 Bikinians to Rongerik Atoll when Bikini Atoll was selected as a ground-zero location for the testing program. Rongerik was too small to support the Bikini population; the Bikinians ravished the islands and the food sources in an effort to survive. By 1948 the Bikinians were in an extreme state of impoverishment due to exhaustion of local food resources and were evacuated from Rongerik and moved to Kili.

During the entire time the Bikinians lived on Rongerik, the U.S. government failed to ask Rongelapese landholders for permission to relocate the Bikinians. When the U.S. government relocated the Bikinians from Rongerik to Kili, it retained Rongerik Atoll for military use, again without the permission of Rongelapese land-holders. Rongerik received extensive fallout during the Bravo event and additional fallout from other weapons tests. No cleanup has ever occurred on Rongerik. The atoll is contaminated and remains off-limits for Marshallese use.

John Anjain: *In 1946 the navy stopped the field trip ship service between Rongerik and Rongelap, and the Bikinians didn't have supplies. Ninety-nine percent of the land on*

---

[5] Tobin, *Land Tenure Patterns,* 29.
[6] Ibid., 31.
[7] John Anjain, record book, 1943, 119.
[8] Ibid., 9.

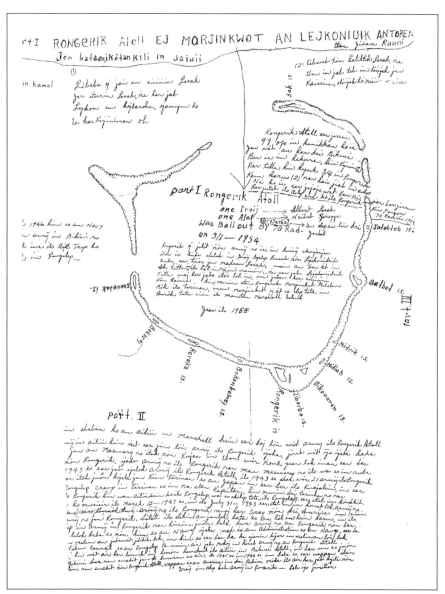

Map 4. John Anjain's map of Rongerik

Rongerik was ruined by the Bikinians. Three-quarters of the land was burned in 1946 by an accidental fire from cigarettes. The people even ate the tiny pandanus and coconut seedlings because the Bikinians were moved from their own homeland and left alone. . . . This made me very sad, because this was the land of my grandfather, and it is difficult to return Rongerik to the way it was before.[9]

As magistrate of Rongelap, John Anjain kept a meticulous record book of events, resources, and actions that influenced the land and lives of the Rongelapese. This book includes documentation of each breadfruit, coconut, and pandanus plant for each weto on Rongerik damaged by the Bikinians, notations that suggest the importance of these resources to the people. The record book also contains a sworn statement asserting that a landowner did not give permission to the United States to make Rongerik available for the Bikinians to use:

Epr. 22-1949. Na Ekinilan W. ij likit ao [text cut off] Ijin bwe iar jab katlok bwe dri Bikini ro Ren komakot jabrewot ion ailin eo ao Rongerik ejelok juon iar jiron kin [text cut off]. Na Ekinilan. W. Dri kamol John Anjain.[10]

[Apr. 22-1949. I, Ekinilan W., I am stating (text cut off) here because I did not give permission for the Bikinians to move anything on my island of Rongerik and no one ever asked me (text cut off). Signed, Ekinilan W. Written (and) witnessed by John Anjain.][11]

SUMMARY OF CONSEQUENTIAL DAMAGES

The U.S. Navy seized, without asking permission from or providing compensation to landowners, *morjinkōt* land owned by the people of Rongelap. The U.S. Navy, also without compensation or permission, placed the people of Bikini on Rongerik Atoll. The people Rongerik never received compensation or other remediation for subsequent damage to Rongerik Atoll by the people of Bikini. Beyond damage inflicted by the Bikinians, Rongerik Atoll was severely contaminated as a result of the Bravo event and other weapons test fallout. The people of Rongelap have been involuntarily displaced from access to and safe use of their customary lands and resources on Rongerik Atoll.

[9] John Anjain, interview by Holly M. Barker, Ebeye, March 16, 1999.
[10] John Anjain, record book, 1949, 78.
[11] Ibid. Translated by Holly M. Barker, Tina Stege, and Tieta Thomas, April 1999.

The Bravo Event

The Bravo event of March 1, 1954, exposed the people and islands of Rongelap, Rongerik, and Ailinginae to dangerous amounts of radiation and other toxic contaminants, such as chemicals used in weapons tests to indicate the amount of fallout from each test. In addition to exposing downwind communities to radiation, "the radioactive debris [from the Castle series] dispersed quickly throughout the world and all but a few stations experienced measurable fallout at sometime during the series."[12] Prior to the Castle series, the U.S. military plan included post-shot evacuation, "if radiological conditions indicate it necessary," on four hours' notice.[13] Weeks before the test, a navy commander visited communities downwind from the test site and "tried to explain something to them about a test, and bombs," but despite the efforts of an interpreter, the people did not understand what he was talking about.[14]

Although the U.S. government had removed the Rongelapese from their homeland for much smaller tests in 1946, no relocation occurred in 1954. Shortly before the Bravo test, the U.S. Department of Interior changed the official "danger zone" and placed Rongelap, Rongerik, and Ailinginae immediately outside the area that would necessitate evacuation, despite the fact that the planned detonation was designed to produce heavy fallout and in fact involved the largest nuclear device ever tested by the U.S. government. The decision to change the boundaries of the danger zone was influenced in part by a February 5, 1953, letter to James P. Davis, director of the Office of Territories at the U.S. Department of Interior, from Elbert D. Thomas, high commissioner of the Trust Territory of the Pacific Islands. High Commissioner Thomas, in an effort to prevent the contamination of critical resources, requested that the danger area exclude Ailinginae. He argued:

> This atoll [Ailinginae], although not "inhabited" in the sense that people are permanently domiciled there, is regularly used and harvested by the people of Rongelap and contributes a substantial part of their living. It is owned by the Iroij of the area and, while it is geographically a separate atoll, it is politically, socially, and economically an integral part of Rongelap. These northern atolls of the Marshall Islands are poor. They offer only the most meager living for the inhabitants. The removal of Ailinginae, or even part of it, from the economic orbit of Rongelap

---

[12] D. E. Lynch, *Radioactive Debris from Operation Castle: Worldwide Fallout* (New York: United States Atomic Energy Commission, New York Operations Office, Health and Safety Laboratory, January 21, 1955), 7, http://worf.eh.doe.gov/data/ihp1d/39683e.pdf (accessed December 17, 2007).

[13] Commander, Task Group 7.1, Los Alamos National Laboratory, letter to distribution, 1953, page 5. Subject: General Concept of Operation Castle.

[14] Special Joint Committee Concerning Rongelap and Utirik Atolls, *A Report on Rongelap and Utrik to the Congress of Micronesia Relative to Medical Aspects of the Incident of March 1, 1954—Injury, Examination and Treatment* (Fifth Congress of Micronesia, February 1973), 81, http://worf.eh.doe.gov/data/ihp1a/1427_.pdf (accessed December 17, 2007).

could be expected to result in critically lowered living standards and it might force migration of part of the population to other area[s] to maintain the subsistence balance of the atolls. There is also no question that such an incursion on their lands will result in serious social repercussions in these islands that might readily extend throughout the Marshalls . . .

While it is impossible to predict the exact nature of the reactions, experience has shown that the most probable result would be first, a lowering of morale with a consequent reluctance to fend for themselves, followed by the expectation that the Government would provide their food in return for the land that had been taken. . . . To them, land, and the lagoons that they fish in, are the only assets which they recognize that they can depend up on to feed them. . . . In fact, land and their fishing grounds are the only assets that can assure these islanders freedom from fear and want. [15]

While on paper, the people of Rongelap were now out of the danger zone, this action did not remove them from harm. Rongerik Atoll was approximately 135 nautical miles from Zero Point, Rongelap was 105 nautical miles, and Ailinginae 83 nautical miles. "For several days prior to March 1, 1954, the Air Force weather station on Rongerik noted the upper winds coming from the west and there was talk about the possibility of part of the radioactive cloud being blown directly over the islands."[16] On the eve of the Bravo test, weather reports indicated that atmospheric conditions were getting less favorable. At midnight, just six hours before the shot, weather reports noted that there were "less than favorable winds at 10,000 to 25,000 foot levels." Winds at 20,000 feet were "headed for Rongelap to the east."[17] Despite the fact that the Bravo event would send fallout blasting toward the inhabited atolls of Rongelap, Rongerik, and Ailinginae, the test continued as planned.

The Rongelapese residing on Rongelap and Ailinginae were exposed to what was then estimated to be 200 roentgens of whole-body radiation (recognized then as a potentially lethal dose), with substantially greater amounts taken up by the thyroid. Children were exposed to even greater levels of radiation. More recently, this exposure was recalculated to a significantly higher level by Dr. Hans Behling (as outlined in Table 2 and contextualized with reference to current known health effects).

---

[15] Elbert D. Thomas, high commissioner of the Trust Territory of the Pacific Islands, letter to James P. Davis, director of the Office of Territories at the U.S. Department of Interior, February 2, 1953, http://worf.eh.doe.gov/data/ihp1b/4060_.pdf (accessed October 25, 2007). It appears that in modifying the mapped danger zone to exclude Ailinginae, the high commissioner assumed that detonation would occur only under atmospheric conditions that did not endanger inhabited atolls or their key resource islands.

[16] Defense Nuclear Agency, eyewitness account of the Bravo incident from a U.S. serviceman posted on Rongerik, 1954. On file at the Embassy of the Republic of the Marshall Islands, Washington, DC.

[17] Defense Nuclear Agency, Pacific Command, midnight weather briefing prior to the Bravo test, March 1, 1954, http://worf.eh.doe.gov/data/ihp1d/18828e.pdf (accessed October 25, 2007).

Table 2. Health Effects of Ionizing Radiation

Adult Exposure Doses for the People of Rongelap from the Bravo Event	
Whole-body doses:	300–375 roentgens
Thyroid:	10,000–20,000 rads
Internal doses (other than thyroid):	60–300 rem
Current Known Health Effects of Ionizing Radiation	
5–10 rem:	Changes in blood chemistry; can cause genetic damage
50 rem:	Can alter white blood cells and produce nausea
75–125 rem:	Can produce radiation sickness
400 rem:	Can kill 50 percent of exposed people
500–600+ rem:	Will kill almost all exposed people

Sources: Hans Behling, *Reassessment of Acute Radiation Doses Associated with BRAVO Fallout, Report to the RMI Nuclear Claims Tribunal* (McClean, VA: S. Cohen and Associates, 2000); "Health Effects of Ionizing Radiation," *Citizens for Alternatives to Radioactive Dumping*, http://www.cardnm. org/backfrm_a.html; "Radiation Protection, Health Effects, *U.S. Environmental Protection Agency*, http://www.epa.gov/radiation/understand/health_effects.html#q1.

Exposures in previous tests in the Marshall Islands and the United States led to the development of safety precautions for minimizing exposure to fallout that were referenced in Operation Castle planning documents. Military personnel were given radiation safety information, and installations on Rongerik and elsewhere were built to minimize exposure. U.S. weathermen were ordered not to eat or drink anything after the test, to wear long-sleeve shirts and pants, and to remain inside their military-built shelters to reduce exposure.[18]

The people of Rongelap were not given the same level of information or access to protective shelters, had no idea the U.S. government would test the Bravo weapon on that day, and were unable to take any protective measures. The *Final Report for JTF-7 Radiological Safety, Operation Castle* (spring 1954) notes:

> *The significant fall-out area from large yield shots was a zone on the order of 100 miles wide and 300 miles long. . . . Native populations were not briefed in advance on the general aspects of the operation, to include approximate starting date of the operation, gross phenomena, which would be manifested, possibilities of hazardous conditions requiring evacuation and general native preparations to centralize and anticipate evacuation.*[19]

---

[18] Undated eyewitness accounts of the Bravo incident from Rongerik by U.S. servicemen Giroux and Rollins, stating that it was known that winds were headed toward Rongerik prior to the detonation of Bravo. DOD document on file at the Embassy of the Republic of the Marshall Islands, Washington, DC.

[19] Joint Task Force Seven, *Final Report for JTF-7 Radiological Safety, Operation Castle, Spring 1954* (Washington, DC: Joint Task Force Seven, 1954), 84–85, appendix C, http://worf.eh.doe.gov/ data/ihp1c/0402_a.pdf (accessed October 25, 2007).

U.S. military personnel monitoring weather and fallout on Rongerik were evacuated on March 2, 1954. Marshallese inhabitants were evacuated on March 4, 1954. Evacuation events are described in memorandums for the record, communications records, and Operation Castle reports. Communications, daily logs, and diary entries from weathermen on Rongerik first reported fallout "over 100" at 1543 on March 1, 1954. At 0030 on March 2, 1954, weathermen on Rongerik were instructed to cease all operations immediately and to remain inside their metal-lined concrete bunkers. They were evacuated beginning on March 2 at 1250, with the final detachment departing at 1800. The memorandum for the record shows no reference to the presence of Marshallese inhabitants until the final message sent by U.S. Air Force captain Louis Chrestensen on March 2, 1954, after landing in Kwajalein at 1900:

> *The final message I sent was an urgent recommendation that Rongelap be surveyed as soon as possible with the expectation that any inhabitants would very likely be subjected to the same degree of fallout that had occurred at Rongerik. This recommendation met approval by CTG 7.4.*[20]

The Rongelapese continued to eat food and drink water tainted from radioactive fallout for the days between the Bravo test and their evacuation to Kwajalein. Both the food and water supplies were obviously contaminated with radioactive fallout, but the Rongelapese had no other food and water options available. After the evacuation of the Rongelapese, the U.S. government verified the contamination levels of the food and water that the Rongelapese had consumed before their removal. Four days after the Bravo test, the U.S. government collected water samples on Rongelap that was found to be radioactive at a rate of two to twenty-five times above AEC operational tolerances.[21]

The Rongelapese suffered near-lethal levels of radiation exposure. Fallout particles stuck to the coconut oil that people used in their hair, and stung their eyes. Children played with the fallout ash, and the entire community ate food and drank water contaminated by radioactive fallout. Several hours after the Bravo test, the people began to feel nauseated. Many vomited and exhibited severe flulike symptoms. By evening, people's skin had begun to blister from contact with the fallout ash. After the people were evacuated, radiation lingered in the air and continued to settle on the atolls.

---

[20] Weather Reporting Element, San Francisco, report of radioactive contamination of personnel at Eniwetak Island, Rongerik Atoll, to Commander, Test Services Unit, March 27, 1954 http://worf. eh.doe.gov/data/ihp1d/18828e.pdf (accessed October 25, 2007).
[21] Joint Task Force Seven, *Radiological Survey of Downwind Atolls Contaminated by BRAVO* (Washington, DC: Joint Task Force Seven, 1956).

Nerja Joseph: *When the fallout came . . . I used the well water that was soaplike because of the fallout dust on top of the water. I soaped my head. I put the fallout on my head. My hair fell out. I am the girl in the Brookhaven picture whose hair fell out.*[22]

Aruko Bobo: *Nearly all the people on Rongelap became violently ill [after Bravo]. Most had excruciating headaches and extreme nausea and diarrhea. By the time of our evacuation to Kwajalein, all the parts of my body that had been exposed that morning blistered, and my hair began to fall out in large clumps. I just had to run my fingers through it and would come away with a fistful.*[23]

Jerkan Jenwor: *Our bodies hurt, we were nauseous. I was sick from the first day. I was really sick after the bomb. I had to stay in the house because I was so sick.*[24]

Aruko Bobo: *I was living with my grandparents on Rongelap. . . . On that March morning [of the test], my grandfather woke me while it was still pitch dark to help other girls in the cookhouse. After awhile, we saw flashing lights, then a loud sound of explosion, then strong wind hit. Chickens and roofing tins flying all over made us frightened. I ran to my grandfather and others. Grandmother immediately came out to see what was causing the children to be that frightened. . . . There were four of us [kids who went outside on the reef] . . . three girls and one boy. The clouds were suddenly all around us . . . the sky had the most unusual color, very scary. . . . The color went from bright white to deep red and then a mixture of both with some yellow. We jumped behind big rocks on the reef. . . . The boy decided we should hurry [to the house]. . . . It was the boy who finally pushed us to run . . . the air around us was split open by an awful noise. I cannot describe what it was like. It felt like thunder, but the force from the noise was so strong that we could actually feel it. It was like the air was alive. . . . Everything was crazy. There was a man standing outside the first hut staring at the burning sky. . . . I found my hair was covered with a white powderlike substance. It had no smell and no taste when I tried tasting it.*[25]

Kajim Abija: *I was in my late twenties at the time of Bravo. I was on Ailinginae with my husband, Jenwor Anjain, who has since died. All of the older people, like my husband, have died . . .*

We used all the small islands on Ailinginae. I went to Ailinginae three times, the third time is when the bomb dropped. . . . When it dropped, we saw a light. It [the light] was to help find submarines my husband said [laughs]. The "powder" [fallout] was on the lagoon side. We were looking for birds when the powder fell. The old man was ready to take his

---

[22] Nerja Joseph, interview by Holly M. Barker, Majuro, March 8, 1999.
[23] Aruko Bobo, interview by Holly M. Barker, Ebeye, August 27, 1994.
[24] Jerkan Jenwar, interview by Holly M. Barker, Ebeye, March 17, 1999.
[25] Aruko Bobo, interview by Holly M. Barker, Ebeye, August 27, 1994. This testimony was reviewed by the informant, and minor details were amended. The translated amendments and details were provided by Senator Abacca Anjain-Madison, September 14, 2001.

boat to go and get birds. He told the children not to play there [on the lagoon side]. I was supposed to take care of the kids, but we all passed out on the lagoon side . . .

When they came to take us away from the island later, our clothes had powder on them. We saw the powder, and we said it was something to reduce the poison from the bomb, the old man said. We were happy because we said we wouldn't get as much poison because that thing was a bomb [laughs].

We prepared the underground oven. We were going to split up the birds. I stayed with the kids. We saw the boats coming, and we wondered why the huge boat came. We had no idea what was happening. We only thought about playing around and finishing our cooking of the birds . . .

On Ailinginae, they didn't stop us from eating and drinking after the powder fell. We blew the powder off of our food and ate it. We couldn't take care of each other, even the kids, because we were all sick. We ate sprouted coconut because we were really nauseous, and when we ate, we got even more nauseous . . . [26]

On the boat, we took showers and put on trousers and sailors' underwear. [27]

SUMMARY OF CONSEQUENTIAL DAMAGES

The Bravo event of March 1, 1954, exposed the people and islands of Rongelap, Rongerik, and Ailinginae to dangerous amounts of radiation and other toxic contaminants, such as the chemicals used in weapons tests to indicate the amount of fallout from each test. This exposure was not unanticipated. Once fallout over populated atolls was confirmed, the U.S. government delayed evacuation of native inhabitants. Native inhabitants received greater levels of exposure to fallout than U.S. service personnel because they were not provided with the same level of information, protective shelter, and risk-reduction advice, and because they were evacuated two days later than U.S. service personnel on Rongerik. By the time of their evacuation, the Rongelapese were exhibiting severe flulike symptoms, radiation burns, and loss of hair.

Relocation from Rongelap to Kwajalein in 1954

The sixty-four residents of Rongelap and eighteen residents of Ailinginae were evacuated to Kwajalein on March 4, 1954. Sixteen people were evacuated by plane and sixty-six by boat. People were forced to leave their homes, personal documents, and belongings such as animals, crops, and boats. According to the Rongelapese, they were not given time to collect any possessions. [28]

---

[26] Marshallese often eat sprouted coconut when they are feeling nauseous or sick.
[27] Kajim Abija, interview by Holly M. Barker, Ebeye, March 18, 1999.
[28] Almira Matayoshi, interview by Holly Barker, Honolulu, June 13, 2001.

The evacuation narrative report by G. W. Albin, commanding officer of the USS *Philip*, confirms rapid departure, with possessions limited to a small handbag, and notes that the people had been sick. Given the severity of illness the Rongelapese experienced immediately after Bravo, there was an urgent need to evacuate them from Rongelap, decontaminate them, and provide them with medical care.

The Rongelapese had serious medical consequences from their exposure that required immediate attention, as indicated in the notes of Dr. Thomas Shipman, one of the attending physicians, who observed that people's "blistered and fissured feet" made it painful for them to walk.[29]

Dorothy Amos: *The boat was fast, and we arrived in Kwajalein in just one day. We got off on the dock. The bus took us—we didn't know to where. Four pregnant women and two older people who couldn't move well arrived by airplane. They put a fence around the place where we were on Kwajalein. It was forbidden for others to enter. Only police and doctors could enter. Every day they had us all go down to the lagoon to wash off. The people on Kwajalein, the Americans, helped us with clothes. On Kwajalein, we ate three times a day. We were treated well.*[30]

Kajim Abija: *The children's hair fell out. Also, the skin of the men who were fishing itched badly. Our fingernails turned black.*[31]

Dorothy Amos: *My hair fell out. It was really funny. I could pull out my hair easily from the burns. Baldness. We were really cooked like they set our heads on fire. . . . I had burns on my arms, throat, legs. It's like I was cooked here [points to throat and arms].*[32]

Almira Matayoshi: *I really cried when we were on Kwajalein. When it was time to decontaminate us, they gave us the men's underwear that the navy men wore. The underwear was too small to cover us, and it was completely see-through when we got wet from the hose they sprayed us with. The water from the hose was so strong too! Billiet Edmond was translating and assisting the navy and saw us all standing there naked. We had tears pouring down our faces because we couldn't believe that our custom was being violated so badly. Billiet was related to so many of the women, and it was like our culture was being ripped apart.*[33]

Lijon Eknilang: *When the U.S. government evacuated people on March 3, 1954, the ship first evacuated the people from Rongelap Atoll. After everyone from Rongelap was on*

---

[29] Joint Task Force Seven, *Reports on Evacuation of Natives and Surveys of Several Marshall Island Atolls* (San Francisco: Joint Task Force Seven, April 9, 1954), http://worf.eh.doe.gov/data/ihp1c/0617_a.pdf; Thomas L. Shipman, letter to Dr. Alvin C. Graves, April 12, 1954. Subject: Natives at Kwaj, http://worf.eh.doe.gov/data/ihp1b/4367_.pdf (accessed October 25, 2007).
[30] Dorothy Amos, interview by Holly M. Barker, Ebeye, March 18, 1999.
[31] Kajim Abija, interview by Holly M. Barker, Ebeye, March 16, 1999.
[32] Dorothy Amos, interview by Holly M. Barker, Ebeye, March 18, 1999.
[33] Almira Matayoshi, interview by Holly M. Barker, Honolulu, June 13, 2001.

board, the ship went to Ailinginae to evacuate the residents. The American personnel on the ship told all the people from Ailinginae to take all their clothes off. Men and women—fathers and daughters, mothers and sons, and relatives that it was extremely taboo to disrobe in front of—were forced to stand naked together while the ship's personnel hosed the people down with water. The evacuees from Rongelap had been through the same process and had already dressed by the time the people from Ailinginae were told to disrobe. . . . Still today, some people find it inappropriate for women to show their thighs, especially near their male relatives, but in the 1950s it was extremely taboo for women to show their thighs, let alone their entire naked bodies. Both men and women were trying to cover themselves with a small towel given to each person and with their hands, but they could not hide their nakedness from one another. We didn't understand at the time why we had to do this . . .

There were two Rongelapese who translated for U.S. government representatives following the emergency evacuation of the Rongelapese, Billiet Edmond and Janwod Anjain. Billiet and Janwod were closely related . . . Billiet's father is the younger brother of Janwod's father. After the Rongelapese were evacuated to Kwajalein, U.S. government doctors required the Rongelapese to bathe in the lagoon three times a day in an effort to reduce their exposure to radiation. These daily baths went on for three months. Each day . . . the Rongelapese women rode in a bus from the camp where the Rongelapese were staying to the lagoon where they bathed. The women were instructed to wear just their underwear and a T-shirt on the bus ride to and from the lagoon. The navy clothes didn't fit the women properly, and they couldn't conceal their bodies properly. Billiet and Janwod accompanied them in order to translate. When the bus arrived at the lagoon, the women were told to stand at the water's edge and take their clothes off. Billiet and Janwod translated. Billiet and Janwod were related to most of the women there. In the Marshallese culture there are strict guidelines directing behavior between male and female relatives. Men and women who are related are not able to discuss anything even remotely sexual—even as a nuance—men should not know when their female relatives urinate for fear that they would think about their relatives' body parts, and under no circumstances should they view one another's bodies in anything but full clothing.

In front of Billiet and Janwod, three times a day for three months, the Rongelapese women were told to undress and stand naked at the lagoon's edge. The women would cry from embarrassment and try to cover their genitals with their hands. U.S. government officials, all men, ran Geiger counters up and down the bodies of the naked women both before and after they bathed in the lagoon. Frequently, the Geiger counters would start clicking wildly when taking readings from the hair on the women's heads and from their pubic hair. The U.S. government workers would tell the women to soap their pubic hair again, in front of everyone, before a second reading. Billiet and Janwod tried to avert their eyes whenever possible, but their presence by their naked mothers and sisters was mortifying.[34]

---

[34] Lijon Eknilang, interview by Holly Barker, Majuro, March 28, 2001.

Almira Matayoshi: *The two hardest things for us to talk about are the divisions in our families caused by the bomb and what happened to our bodies. The Rongelapese who weren't exposed wouldn't admit they were Rongelapese. This was awful, because we are family, and this is the worst kind of damage to have splits in the family.*[35]

Isao Eknilang: *We were very isolated on Kwajalein. Our relatives on Ebeye were afraid of their own family members, they were afraid to visit us for fear they would get radiation from us. Even the Rongelapese who were not on Rongelap during Bravo became embarrassed to be Rongelapese. They wouldn't want to admit they were Rongelapese. They were embarrassed because we were like monkeys. Our arrival on Kwajalein caused family divisions, because family did not want to help us for fear of being exposed themselves.*[36]

SUMMARY OF CONSEQUENTIAL DAMAGES

The Marshallese experienced cultural violations and indignities during evacuation and decontamination efforts on the ship and on Kwajalein. When Marshallese on Kwajalein saw that the Rongelapese were sick from radiation exposure, they became afraid to go near or help them, for fear they would contract the same illnesses. These reactions created lasting impressions on family members and started the first serious divisions within families.

Project 4.1 Research on Kwajalein

Throughout the 1940s and 1950s, accidental exposures at research facilities and test sites provided scientists with radiation-effects research opportunities. By the time of the Bravo event in 1954, U.S. researchers were well positioned to take advantage of the scientific opportunities created by human exposure to radiation. Many research projects were underway in 1954. These included studies, originally proposed by the Ad Hoc Committee for Biologic Tests in June 1949, to define the research agenda for "future proof tests of atomic weapons at Eniwetok."[37]

Per recommendations to the U.S. Atomic Energy Commission, outlined in a 1953 review of research needs by the advisory committee, the Bravo-event exposure of U.S. service personnel and Marshallese natives prompted immediate efforts to collect urine and blood in support of ongoing biomedical research on the effects of radiation on a human population. This research included documentation of "the

---

[35] Almira Matayoshi, interview by Holly M. Barker, Honolulu, June 13, 2001.
[36] Isao Eknilang, interview by Holly M. Barker, Honolulu, June 13, 2001.
[37] Ad Hoc Committee for Planning Biological Aspects of Future Atomic Bomb Tests, letter to Dr. Shields Warren, Divison of Biology and Medicine, U.S. Atomic Energy Commission, June 10, 1949), http://worf.eh.doe.gov/data/ihp1b/4075_.pdf (accessed October 25, 2007).

carcinogenic action of ingested or inhaled radioactive materials."[38] After evacuating the people of Rongelap, scientists returned to the atolls to take samples of water and vegetation and to capture exposed animals. Sixty-six animals (chickens, pigs, cats, ducks) were brought back for study, since "if anything does show up it will be more likely to show up sooner in the animals, and would give us some idea of the prognosis for the humans over a longer period of time."[39] In addition to changes in white cell and platelet levels, radioautographs of chicken and pig tibiae showed abnormal bone morphology. Samples from two pigs indicated "two separate and distinct exposures to fallout material" . . . "most likely strontium and barium."[40] This evidence indicates that the Bravo event was not the first weapons test to produce fallout on Rongelap.

On March 9, 1954, Project 4.1 scientists arrived on Kwajalein. On March 11 the Rongelapese began service as subjects in a variety of studies exploring the effects of radiation exposure on human beings. Over the three months following their evacuation, 64 residents of Rongelap, 18 residents of Ailinginae, 157 residents of Utrik, and 28 Americans from Rongerik—together with control groups of 117 Marshallese living in Majuro and 105 U.S. service personnel—had their conditions monitored and emerging symptoms documented in support of Project 4.1.

Project 4.1 activities on Kwajalein are described in report documents as including efforts to (1) document and treat immediate effects from acute radiation exposure; (2) document the population and control groups in ways that set a baseline for further studies on the long-term effects of radiation; (3) obtain samples, measurements, and biological responses that suggested exposure levels, and (4) provide information to ongoing studies on absorption rates, elimination processes, and other questions of interest to the national security and military defense of the United States. A report was produced from data collected from March to May 1954, submitted to the AEC Division of Biology and Medicine for discussion at its July 12–13 Conference on Long Term Surveys and Studies of the Marshall Islands, and formally released in October 1954.[41] This final report summarized findings from

---

[38] General Advisory Committee to the U.S. Atomic Energy Commission, minutes of thirty-sixth meeting, August 17–19, 1953, 14–16.

[39] S. H. Cohn, *Conference on Long Term Surveys and Studies of the Marshall Islands* (Washington, DC: AEC Division of Biology and Medicine, 1954), 59, http://worf.eh.doe.gov/data/ihp1c/0246_a.pdf (accessed October 25, 2007).

[40] Ibid., 60–61.

[41] E. P. Cronkite, V. P. Bond, L. E. Browning, W. H. Chapman, S. H. Cohn, *Study of Response of Human Beings Accidentally Exposed to Significant Fallout Radiation* (Bethesda, MD: Naval Medical Research Institute, October 1954). See also Cohn, *Conference on Long Term Surveys;* and Thomas Shipman, Health Division, Los Alamos Scientific Laboratory, telex to John Bugher, AEC Division of Biology and Medicine, March 10, 1954, http://worf.eh.doe.gov/data/ihp1d/400045e.pdf (accessed December 17, 2007).

"secret restricted data" developed from "a joint AEC-DOD [study] established to study the physiological symptoms of evacuated natives."[42]

Project 4.1—the Study of Response of Human Beings Exposed to Significant Beta and Gamma Radiation Due to Fallout from High Yield Weapons—"represented the first observations by Americans on human beings exposed to excessive doses of radiation from fallout."[43] Conveniently, "the dosage spread of the different groups nicely cover[ed] the range of estimated operational tolerance accepted by the Department of Defense."[44] Project 4.1 was thought to be a valid research study for the following reasons: "The groups of exposed individuals were sufficiently large to provide good statistics . . . the exposures involved far exceed the normal permissible dosage. . . . The internal dosage was due mostly to ingested material rather than inhaled material . . . [and] beta activity in the urine of these exposed human beings indicated significant internal contamination."[45]

Loss of hair, depressed blood cell and leukocyte counts, flulike symptoms, fingernail discoloration, nausea, and radioisotope activity in the urine were all observed in the Rongelapese following their acute external exposure.[46] The skin burns experienced by the Rongelapese were of great interest. By examining the exposed Rongelapese, researchers found that beta radiation from fallout penetrate well into the body with a large portion being absorbed by the critical living layers of skin. And they were able to equate the observed degree of skin damage suffered by the Marshallese natives during Operation Castle to known dose-effect data,

---

[42] As described in *Operation Castle: Report of the Manager, Santa Fe Operations,* Spring 1954. On file at the Embassy of the Republic of the Marshall Islands, Washington, DC.

[43] Armed Forces Special Weapons Project, *Operation Castle: Summary Report of the Commander* (Albuquerque, NM: Armed Forces Special Weapons Project, January 30, 1959), 71.

[44] *Operation Castle, Pacific Proving Grounds, Joint Task Force-7, Report of Commander Task Group 7.4,* April 1954. On file at the Embassy of the Republic of the Marshall Islands, Washington, DC.

[45] Armed Forces Special Weapons Project, *Operation Castle,* 71–72. Over the years, questions have been raised as to whether the Bravo event was a purposeful exposure of the Rongelap people, thus creating a suitable population for study on the long-term effects of radiation (see, for example, the Dennis O'Rourke film *Half Life*). An ACHRE investigation into this question found no evidence to support this contention. However, the investigation was limited to declassified materials made available in 1994 and early 1995. Documents declassified in the ensuing years continue to raise questions. See, for example, D. Curry, Los Alamos National Laboratories, Joint Task Force Seven, Task Group 7.1, memo to A. C. Graves, November 10, 1953. Subject: Outline of Scientific Programs—Operation Castle. Document J-2136, page 2. Included in the list of research programs is Program 4, Bio-Medical Studies, directed by Commander E. P. Cronkite, with Project 4.1 described as "The Study of Response of Human Beings Exposed to Significant Beta and Gamma Radiation Due to Fall-out from High Yield Weapons." A note on the document states, "attached as separate entry."

[46] *Issues Affecting United States Territory and Insular Policy* (Washington, DC: United States Government Accounting Office, 1955); *Operation Castle: Summary Report of the Commander;* Joint Task Force Seven, *Final Report for JTF-7 Radiological Safety,* 84–85, appendix C.

concluding that beta burns will occur at sublethal gamma levels only when particles come into contact with bare skin.[47]

William Allen (who accompanied U.S. representatives on surveys of the islands): On one of these trips I met a Marshallese male named Hiroshi who had been severely affected by fallout. He had first-degree burns covering 90 percent of his body and had suffered complete loss of hair. The extent to which his body was burned was such that the bones in his feet were exposed and visible to the naked eye. Tragically enough, Hiroshi died less than a year after our conversation.[48]

Researchers noted that the severity of skin burns experienced by the Rongelapese may have been exacerbated by chemical irritants present in the fallout. A U.S. government document notes that irritating chemicals applied to the skin during or shortly after irradiation enhance the effects of radiation and that the chemical nature of the fallout may have enhanced the effects of radiation on the Rongelapese.[49]

During Project 4.1, U.S. researchers did not obtain informed consent from Rongelapese subjects for exams, procedures, and sampling. Nor did they explain the procedures. Approval for research was sought and secured through U.S. military and scientific advisory committees. The territorial administration was notified, and it approved requests to collect urine samples in support of a preexisting study of plutonium secretion (a research project established after accidental exposure of scientists at Los Alamos).[50] The people of Rongelap, however, were not involved in these discussions.

Chiyoko Tamayose: *We never knew what was going on. There was a time where they took my blood, mixed it with something, and then shot it back into me. They never asked me if they could do this, they just did it. I didn't understand what they were doing, and I still don't.*[51]

[47] Robert A. Conard, Eugene P. Cronkite, Victor P. Bond, James S. Robertson, and Stanton H. Cohn, *Fallout Radiation: Effects on Marshallese People* (Upton, NY: Brookhaven National Laboratory), http://worf.eh.doe.gov/data/ihp1a/1030_.pdf (accessed April 1, 2008).
[48] William Allen, interview by Holly M. Barker, Majuro, August 22, 1994. Hiroshi accompanied Aruko Bobo across the reef, see testimony for note 157. Hiroshi is also depicted in photo 12.
[49] Cronkite et al., *Study of Response of Human Beings,* part 1, page 231; part 2, page 615. The U.S. government has yet to disclose the nature of these radiation-enhancing chemicals, their persistence in the environment, or their effects on human beings.
[50] "We have local approval . . . to collect and analyze urine samples from local natives and air weather service personnel," Los Alamos Scientific Labortory, telex to U.S. Atomic Energy Commission, Washington, DC, http://worf.eh.doe.gov/data/ihp1d/400045e.pdf (accessed October 25, 2007).
[51] Chiyoko Tamayose, interview by Holly M. Barker, Honolulu, June 13, 2001.

SUMMARY OF CONSEQUENTIAL DAMAGES

Without their consent or knowledge, on March 3, 1954, the Rongelapese were enrolled in a secret U.S. military project to study the effects of radiation exposure on human beings. The U.S. government used Marshallese blood, urine, bone marrow, thyroid glands, and bodies to calculate how much radiation had been ingested from the environment following the Bravo event, to determine immediate effects of radiation exposure, and to predict long-term effects. Forced and unexplained medical procedures, pain and suffering from injuries and procedures, humiliation from examination and decontamination processes, humiliation from being photographed naked, and social stigmatization by nonexposed relatives were some of the many consequences of involvement in initial Project 4.1 exams and experiments.

Relocation from Kwajalein to Ejit

At the conclusion of three months of short term-studies for Project 4.1 exams and experiments, the Rongelapese were moved from Kwajalein to Ejit, where they were told they would stay for one year until their islands were safe to return to. They stayed on Ejit from June 1954 until May 1957. The move to Ejit was documented by military films and journalists to placate global concerns following the Marshallese filing of a petition with the United Nations requesting assistance in ending nuclear weapons tests. U.S. government reports of the relocation of the Rongelapese from Kwajalein to Ejit Island on Majuro describe the relocation as "a modern version of the American covered wagons," as "natives" were transferred on covered ships loaded with their personal possessions, children, and household items.[52] Declassified documents describe this move as allowing the continued scientific study of the exposed population and recognized that Rongelap, Rongerik, and Ailinginae remained too dangerous for human resettlement. Resettlement was consciously designed to meet bare minimal needs of the Rongelapese, as "it is the policy of TERPACIS [Trust Territory of the Pacific Islands] . . . to discourage too rapid acquisition of wealth by small groups of natives . . . thus, the subsidy of natives is to be held to the essential minimum."[53]

The people of Rongelap lived on Ejit until their return to Rongelap in May 1957. During this time, they encountered hardships associated with adjusting to new surroundings, inadequate access to critical resources, and anxieties over their

---

[52] Cronkite et al., *Study of Response of Human Beings,* part 1, page 231; part 2, page 435.
[53] U.S. Army colonel David O. Byars Jr., report to Commander in Chief, Pacific, April 30, 1954, 5. Subject: Survey of Rongelap and Utirik Atolls, http://worf.eh.doe.gov/data/ihp2/7636_.pdf (accessed October 25, 2007). TERPACIS refers to the Trust Territory of the Pacific Islands.

medical and living conditions and their future. They were especially anxious about radiological conditions on Rongelap, fearing that the fallout that forced their initial evacuation might cause them to suffer the same fate as the dislocated Bikinians.

Almira Matayoshi: *We were on Ejit. . . . All that time, we didn't have school. We didn't have knowledge from school, and we didn't learn from our parents and grandparents about how to weave and make things the way we did on Rongelap.*[54]

Norio Kebenli: *When the community was on Kwajalein after Bravo, they were fed very well and didn't have to cook. . . . Suddenly they were moved to Ejit, with no cooking utensils or appliance[s] . . .*

They used the cardboard boxes from the C-rations and K-rations to build mon tutu *(a separate bathing area) and to paper the walls of their shelters to prevent rain from seeping through. Houses were studio type, and it was very hard.*[55]

On Ejit there was not enough water. The groundwater was bad for drinking. We went all over the island digging holes and trying to find good water, but there was none . . .

It was while on Ejit that life started to get hard. There wasn't enough food; there was no breadfruit and no place to fish . . .

We also began to get some bad habits on Ejit, like learning to smoke from the navy guys. We would also throw away our litter on the beach because we were used to throwing our food garbage on the ground. On Ejit, kids didn't know to be careful about garbage. I remember that several kids cut themselves on the sharp lids from the food the U.S. gave us in cans. These cuts were really deep . . .

We also tried sweet foods for the first time. We were given C-rations and K-rations. We liked the K-rations better because they had sweet foods in them. We used to eat the sweet stuff and throw out the rest.[56]

SUMMARY OF CONSEQUENTIAL DAMAGES

The people of Rongelap endured hardships from food and water shortages on Ejit, as well as anxiety about not knowing when they could return home. Social and cultural assaults included loss of opportunity to practice subsistence-oriented production, the loss of three years of formal schooling opportunities, and exposure to U.S. consumer lifestyles and foods.

---

[54] Almira Matayoshi, interview by Holly M. Barker, Honolulu, June 13, 2001.
[55] This testimony was reviewed by the informant, and additional detail was added. The detail was translated and submitted to the authors by Senator Abacca Anjain-Maddison, September 14, 2001.
[56] Norio Kebenli, interview by Holly M. Barker, Honolulu, June 13, 2001.

Long-Term Human Subject Research Plans, Priorities, and Policies

The first known experiment involving the use of Marshallese human subjects was funded by the U.S. Atomic Energy Commission in 1951. Geneticist James V. Neel received three research contracts to work with a Marshall Islands population, examining naturally occurring mutations in a human population. This research was expected to "yield data of considerable value in our thinking about the genetic effects of irradiation in man."[57] Dr. Neel's expectation with this research was "that the spontaneous rate of mutation in human genes is going to be relatively high, a fact which of course would make their apparent sensitivity to irradiation of less relative importance in the overall picture of man's decline."[58]

Experimental procedures, exams, and samples collected from March through May 1954 in support of Project 4.1 represent the next known use of Marshallese human subjects by AEC scientists, and it is clear from study documents that findings from the initial human radiation study were merely the first reports in a long series of anticipated studies. The initial Project 4.1 report concluded: "It is not possible at this time to give an estimate of the long term effects of internal contamination. . . . It is expected that additional data will be available in the near future that will make possible reasonable estimates of the long term effects of internal radiation on these individuals."[59] Concerning the health and future of the people of Rongelap,

---

[57] Scientific Advisory Committee of the Atomic Energy Commission, *Monthly Status and Progress Report for February 1951* (Washington, DC: Atomic Energy Commission, 1951), http://worf.eh.doe.gov/data/ihp1b/7717_.pdf (accessed December 17, 2007).

[58] James V. Neel, letter to Dr. Max Zelle, Biology Branch, Division of Biology and Medicine, U.S. Atomic Energy Commission, February 20, 1951. ABCC box 19, genetics (3B), 1949–1955, National Academy of Sciences Archives. While this research is the first incident of human subject work involving the Marshallese, efforts to locate the formal research protocol, field journals, trip reports, or other data relating to this project have been fruitless. Neel's research is described in AEC committee minutes and press releases as representing the crucial first step to understanding the genetic effects of radiation in a human population. Difficulties in locating documentation on the fieldwork may, perhaps, be explained by the fact that this research did not involve the use of radiation; thus information relating to this experiment would not have been declassified under the Clinton administration order. Neel's 1951 description of his plans for Marshallese research suggest that it involved the introduction of a naturally occurring agent (possibly a viral vaccine) to document mutagenic response in blood proteins. Thus blood samples would have been collected before and after inoculations. These data might have been used to establish a genetic baseline of a Marshallese population. Neel also requested additional samples from Rongelapese subjects in support of his genetics research in a September 25, 1957, letter to Dr. Robert Conard (http://worf.eh.doe.gov/data/ihp1b/3797_.pdf). The 1958 survey (http://worf.eh.doe.gov/data/ihp1a/4569_.pdf) indicates that samples were obtained. Neel, Robert Ferrell, and Robert Conard eventually collaborated on a formal publication; see James V. Neel, Robert Ferrell, and Robert Conard, "The Frequency of "Rare" Protein Variants in Marshall Islanders and other Micronesian Populations," *American Journal of Human Genetics* 28, no. 3 (May 1976): 262–69.

[59] Cronkite et al., *Study of Response of Human Beings,* part 1, page 231; part 2, page 263.

Project 4.1 recommendations included the warning that the people "should be exposed to no further radiation, external or internal with the exception of essential diagnostic and therapeutic x-rays for at least 12 years. If allowance is made for unknown effects of surface dose and internal deposition there probably should be no exposure for [the] rest of [their] natural lives."[60]

The AEC Division of Biology and Medicine discussed data and findings from Project 4.1 at the July 1954 Conference on Long Term Surveys and Studies of Marshall Islands and used this data to shape the third major human subject research event—the formation of an integrated long-term human environmental research program to document the bioaccumulation of fallout and the human effects of this exposure.[61] From the start, it was assumed that the people of Rongelap would be returned to a contaminated setting and that the recommendation for no further exposure for at least twelve years would be ignored. The AEC Division of Biology and Medicine Advisory Committee discussion included research needs and the logistics of studying the uptake of fission products in the food chain before and after returning the people of Rongelap to their atoll:

Mr. Harris: *These people will introduce animals when they get back there. There may be a concentration of certain isotopes, such as strontium 90, for instance, in these animals or in the meat parts of these animals, which would subsequently be eaten by the natives. If you put animals back, immediately you would get an idea at measurable levels of what this translocation rate might be. Remembering that people are going to eat these animals later on.*[62]

Dr. Bugher: *In this consideration, how do you feel about simply taking specimens at intervals as the islands are visited from their pigs, from their chickens and dogs, and from the people as they die too, if you can possibly get the material . . . wild animals might have been mentioned here . . . the shell fish and crabs. . . . In the surveys, specimen material of these various indigenous fauna would be desired to these various groups concerned with these analyses.*[63]

To allow long-term monitoring of the degenerative effects of radiation, regular interval examinations were proposed. These consisted of:

---

[60] Eugene Cronkite, memorandum to the Commander of Joint Task Force Seven regarding "Care and Disposition of Rongelap Natives," April 21, 1954, http://worf.eh.doe.gov/data/ihp1b/7576_.pdf (accessed April 1, 2008).
[61] Cohn, *Conference on Long Term Surveys,* 59.
[62] Ibid., 220.
[63] Ibid., 223–24.

- Physical examinations and interval histories
- Hematological studies such as hematocrit, white-blood-cell count, differential count, platelet enumeration, and bone marrow studies
- Full-body skin exams documented in color and black-and-white photographs, with skin biopsies if indicated
- Ophthalmological studies of the lens, including a photographic record of the anterior portion of the lens
- Growth studies of children, including the development of dentition
- Studies of the progress of pregnancies and status of newborn infants. (In discussing this proposal, Bugher remarks, "I don't think much comment is required except that essentially it will be a documentation of nothing happening in all likelihood.")[64]
- Quantitative studies of internally deposited radioisotopes by means of urinary-excretion measurements, external radiological measurement and localization, and such radiography as may be useful
- Environmental surveys of the affected islands and atolls and appropriate examination of the animals left on contaminated islands.

In addition to identifying the focus of long-term research, the advisory committee identified those areas it was not interested in pursuing. With reference to questions of fertility, while "abnormal menses were observed in two women in the Rongelap group" and fertility studies on animals brought back from Rongelap suggested a 50 percent fertility rate,[65] scientists balked at the prospect of research examining human reproductive and intergenerational effects of radiation:

Cdr. Cronkite: *My feeling toward it is very simple. We should not attempt to do any studies for fertility for obvious psychological reasons for natives themselves. It becomes a fairly personal thing for getting specimens of semen and prying into these things. It is difficult enough to get a specimen of urine, and feces, let alone inducing masturbation on a large scale of Marshallese.*

Dr. Dunham: *If properly induced. You don't know who the fathers are. You are dealing with a group where there is no control. You would have to use the Uterikans as control. . . . Furthermore, the data in Japan suggests that as far as live births are concerned, there are pretty good data on that. A lot of it where large numbers of people studied both control and irradiated population, and there is apparently no difference.*[66]

---

[64] Ibid., 146.
[65] Ibid., 54.
[66] Ibid., 106–07.

Agreements emerging from this conference included recognition that the "type of study which was made in the acute phase will need to be continued for an indefinite time, but with a changing emphasis from what might be called acute problems to the long term effects which are particularly likely to manifest themselves in such things as shortening of life, the occurrence of tumors, both superficial and deep, and in bone changes, which may be minor in nature."[67] U.S. researchers, therefore, monitored the Rongelapese with the expectation of seeing the people contract these conditions.

Interval examinations were also recognized as opportunities to collect samples of interest to a wide range of ongoing studies:

> **Dr. Bugher**: *Now, our feeling was, too, that various groups of people have special interests and would like to have sample material of various kinds. As far as possible, the groups concerned with the interval study should attempt to provide those samples.*[68]

Some of those with "special interests" included researchers seeking teeth extracted from subjects to support ongoing studies on the effects of radiation on growing teeth, as well as those seeking a selection of teeth extracted during during autopsies.[69] Others wanted bone marrow aspirations to support hematological studies and the use of radioisotopes in tagging procedures to measure red-blood-cell formation rates.[70] Yet others wanted interval x-rays on long bones of living subjects to track roentogenographic changes and to study radio element deposition, and the harvesting of various samples in postmortem autopsies.[71]

The sampling of human subject material that continued after death violated Marshallese traditions, as noted in a May 21, 1956, letter from the district administrator of the Trust Territory of the Pacific Islands to Dr. Robert Conard, Brookhaven National Laboratory:

> *Jabwe, the health aide on Ejit came over to Uliga early in the evening to announce that one man had suddenly died after a very short illness of about one hour duration. Since this was the first death among the original Rongelapese that were brought here, we were faced with the problem of an autopsy. Not to interfere with local customs of Marshallese death rituals we agreed to let the family have the body for the night. Mr. Bender and Mr. Jack Tobin [anthropologists working with the TTPI] had asked the family permission for an autopsy but the family refused. The family were reluctant with a fear that we shall mutilate the body,*

[67] Ibid., 242.
[68] Ibid., 147.
[69] Ibid., 109, 111.
[70] Ibid., 174–75.
[71] Ibid., 183–84; 189–90.

remove pieces and organs which according to their native belief is a very bad thing to the future afterlife of the deceased.[72]

Because long-term research plans involved an integrated study focusing on the absorption and effects of radioisotopes in ecological and human systems, medical research plans paralleled ecological survey efforts. Rongelap was resurveyed beginning in 1954, with soil and food samples taken from Rongelap Atoll by biologists, geologists, and other scientists to learn about the movements of isotopes in the environment.[73] In 1954 U.S. researchers identified cesium-137 (Cs-137) as "one of the principal radio nuclides found at Rongelap."[74] Cesium-137 was detected in edible portions of plants prior to the resettlement of the Rongelapese; the average counts in soils for Cs-137 appeared higher at Rongelap than at Bikini or Enewetak.[75] A 1955 survey conducted by the University of Washington on Rongelap and Ailinginae indicated that "edible plants other than coconuts, such as pandanus, papaya, and squash, have been found to contain levels of Sr-90 which are above the tolerance level as defined in the Radiological Health handbook."[76] In general, plants in the north had higher levels of radiation than those on islands in the south, with the exception of arrowroot, a staple of the Rongelapese. Arrowroot on Rongelap Island was found to have almost three times more radiation than arrowroot in other sample areas on Rongelap and Ailinginae in 1955. The increased radiation levels of arrowroot on Rongelap Island were attributed to the fact that the arrowroot at Rongelap was probably collected in radiation hot spots. Unfortunately for the Rongelapese, these hot-spot "readings were highest in soil depressions and in pits such as those used by the natives for growing crops."[77]

Researchers also found that coconut crabs fed on land plants with high levels of strontium-90 (Sr-90) and Cs-137.[78] In a 1955 survey of Rongelap and Ailinginae, radiation ecology studies indicated that "the highest Cs-137 levels were found in

---

[72] District administrator of the Trust Territory of the Pacific Islands, letter to Dr. Robert Conard, Brookhaven National Laboratory, May 21, 1956. On file at the Republic of the Marshall Islands Embassy, Washington, DC.

[73] Joint Task Force Seven, *Final Report for JTF-7 Radiological Safety*, 84–85, appendix C. See also, "Project Hardy (The Return of the Native)" (research proposal, Los Alamos National Labortory, April 14, 1954), http://worf.eh.doe.gov/data/ihp1c/0607_a.pdf (accessed April 1, 2008).

[74] Lauren R. Donaldson, *A Radiological Study of Rongelap Atoll, Marshall Islands, During 1954–1955* (Seattle: University of Washington, 1955), 164.

[75] H. V. Weis, S. H. Cohn, W. H. Shipman, J. K. Gong, *Residual Contamination of Plants, Animals, Soil, and Water of the Marshall Islands Two Years Following Operation Castle Fallout* (San Francisco: United States Naval Radiological Defense Laboratory, 1956).

[76] Staff of the Applied Fisheries Laboratory, *Radiobiological Resurvey of Rongelap and Ailinginae Atolls Marshall Islands October–November 1955* (Seattle: University of Washington, December 30, 1955), 31–32, http://worf.eh.doe.gov/data/ihp1c/0696_a.pdf (accessed December 17, 2007).

[77] Ibid., 31.

[78] Ibid., 48.

the land plants and the coconut crab," accounting for 26 to 100 percent of the radio-activity in the specimens that year.[79] Because researchers understood that coconut crabs bioaccumulate radiation, one year before the Rongelapese returned home, a recommendation was made to resettle the people provided that they did not eat coconut crab.[80] This recommendation was not implemented until several years after the people returned. The resulting exposure allowed scientists to measure the amount of radionuclides present in human beings and confirm the levels of fallout produced by the weapons, since "the deposition in the human body seems roughly to parallel the levels of fallout. More data are necessary to validate this point fully."[81]

In 1957 the U.S. government announced to the Rongelapese that it was safe to return to Rongelap. This assurance was based on assumptions that initial high levels of radioactivity had subsided and that the levels detected from airplane readings at 200 feet in 1957 were the same as levels present on the ground.[82]

The decision to repatriate the people of Rongelap to a still-contaminated setting ignored medical recommendations to avoid future exposures. This decision also supported U.S. scientific research and military defense agendas. In 1957 the U.S. government was battling an effort at the United Nations to produce a test ban treaty. Given global concern over the plight of the Rongelapese, the U.S. government worried that any future evacuation to protect the population would generate negative publicity, highlighting the fallout dangers of nuclear weapons tests. Injuries from the Bravo event had been highly publicized, and U.S. officials worried that further injury might produce a resolution from the United Nations banning

---

[79] Ibid., 58.

[80] Staff of the Applied Fisheries Laboratory, *Radiobiological Resurvey of Rongelap and Ailinginae,* 58; Edward E. Held, *Land Crabs and Radioactive Fallout at Eniwetok Atoll* (Seattle: University of Washington Applied Fisheries Laboratory, May 27, 1957); Gordon Dunning, *Return of Rongelapese to Their Home Island* (Washington, DC: Atomic Energy Commission, February 6, 1957). The AEC report observes, "Several radiological surveys of the Marshall Islands, especially Rongelap Atoll, have been made during the past two and one-half years. The latest survey (July 23–24, 1956) indicates a presence of a residual contamination on the island of Rongelap, but at a level that is acceptable from a health point of view both for the potential external gamma radiation exposure and the strontium-90 content in the food supply, with the possible exception of land crabs. Therefore, it is recommended that the position of the Atomic Energy Commission should be that the Rongelapese could be returned to their home island as soon as rehabilitation procedures are completed, with the advice that land crabs not be eaten at this time." See also, Dunning, *Radioactive Contamination of Certain Areas in the Pacific,* 50–51.

[81] W. F. Libby, "Current Research Findings on Radioactive Fallout," *Proceedings of the National Academy of Sciences: Physics* 42 (1956): 953.

[82] Gordon Dunning criticized the decision to return the community to Rongelap without adequate radiological assessment. He noted in a letter to the director of the Division of Biology: "I had assumed there would be another radiological survey of the Rongelap atoll just prior to the return of the Rongelapese. . . . It would have been highly preferable to have had a complete survey of the atoll, especially the foodstuffs, but it appears we will have to settle for the external

future weapons tests. These concerns are reflected in a discussion of the committee charged with making the repatriation decision—the AEC Advisory Committee on Biology and Medicine (ACBM):

> *The current low morale of the natives was pointed out and the advantages of returning them to their homes presented as a factor which should be balanced against the possible radiation hazard in their return. . . . It was agreed that because of the already high relative exposure to which the natives had already been subjected, limiting their exposure in terms from now on was unrealistic; but on the other hand, the psychological effect of permitting them to receive more radiation than our own people, could be subject to criticism.*

Further discussion resulted in adoption of a formal statement expressing the committee's opinion:

> *It is moved that the ACBM approve the Division of Biology and Medicine's proposal to return the Rongelapese to their native atoll. However, it is the opinion of the ACBM that if it should become necessary to re-evacuate because of further tests there would result world opinion unfavorable to the continuation of weapons testing.*[83]

The ACBM statement in effect produced a nonevacuation policy, by which the Rongelapese were returned to Rongelap with the understanding that they would remain there regardless of the potential for hazardous exposure in future tests. As a result, the resettled Rongelapese— exposed and "nonexposed"—endured the ill effects from cumulative exposure to multiple weapons tests.

In July 1957 a LCU (landing-craft unit) transported the Rongelapese from Ejit to Rongelap Island, where some effort had been made to construct prefabricated housing and related improvements.[84] By U.S. naval edict, use of Rongelap Atoll was restricted to just Rongelap Island. Similarly, Ailinginae Atoll was off-limits. The Rongelapese were not allowed to return to Rongerik because the environment still

---

readings only." Gordon M. Dunning, letter to C. L. Dunham, director of the Division of Biology and Medicine, July 13, 1957. Subject: Regarding a resurvey of Rongelap Atoll (reproduced in the appendix to this report); Gordon M. Dunning, memo to Commander P. F. Bankhardt, United States Navy, October 1, 1957. Subject: Regarding a resurvey of Enaetok Island. On file at the Embassy of the Republic of the Marshall Islands; Gordon M. Dunning, letter to Dr. A. H. Seymour, Environmental Sciences Branch, Division of Biology and Medicine, February 13, 1958. Subject: Regarding operational responsibilities for the resurvey of Rongelap, http://worf. eh.doe.gov/data/ihp1d/400209e.pdf (accessed April 1, 2008).

[83] Advisory Committee on Biology and Medicine, meeting minutes, November 16–17, 1956; G. Faila, Advisory Committee on Biology and Medicine, letter to Lewis L. Strauss, U.S. Atomic Energy Commission, November 19, 1956, 2. The letter to Strauss is reprinted in the appendix to this report.

[84] Jack A. Tobin, *Report of Repatriation of the Rongelap People for US AEC* (Albuquerque, NM: Holmes and Narver, November 1, 1957). On file at the Nuclear Claims Tribunal, Majuro, RMI.

had not recovered from the Bikinian depletion of the resource base, and because it remained contaminated from the tests. As a result, the resettled Rongelapese stopped using Rongerik Atoll.

Shortly after the Rongelapese returned, scientists went to Rongelap to collect foodstuff samples and analyze them for radiological contamination. The U.S. military wanted to conduct nuclear weapons pretest surveys of Rongelap to determine residual amounts of radioactivity before conducting another series of nuclear tests (Hardtack Series 1 in 1958). A July 1957 survey of food, soil, water, and plankton indicated that radiation levels were "appreciably above background."[85]

The nonevacuation policy meant that the people of Rongelap had to contend with contamination from the Bravo event, pre-1954 tests, and tests that had occurred during their absence, such as the 1956 Redwing series. And, because of the policy of no reevacuation, they faced additional contamination from later military operations. For example, the 1958 Operation Hardtack series, with its detonation of thirty-three nuclear weapons on Bikini and Enewetak, generated additional fallout on Rongelap, producing elevated levels of plutonium noted in subsequent medical and environmental surveys.[86] Research conducted by University of Washington scientists in 1961–1962 documented increased concentrations of radioactive iron (Fe-55) in goatfish liver, and medical survey scientists documented subsequent increases in average Rongelapese body burdens of Fe-55.[87]

And reports published in 1999 indicate that in the summer of 1968, the Deseret Test Center conducted a series of tests known as DTC 68-50 from the USS *Granville S. Hall,* anchored off Enewetak Atoll. This test series involved the atmospheric dissemination of "PG"— staphylococcal enterotoxin B—a toxin that causes incapacitating food poisoning, with flulike symptoms that can be fatal to the very young, the elderly, and people weakened by long-term illness. Staphylococcal enterotoxin B was disseminated over a 40- to 50-kilometer downwind grid, and according to the final report, a single weapon was calculated to have covered 2,400 square kilometers, an area equal to 926.5 square miles.[88] The medical survey for

---

[85] Joint Task Force Seven, *Operation Hardtack: Report to the Scientific Director* (Joint Task Force Seven, August 1, 1958), 57.

[86] L. R. Donaldson, letter to A. H. Seymour, January 11, 1957. Subject: Difference in Activity Levels of the Water at Rongelap as Reported in Earlier Visits Compared with a Visit Done in 1956, http://worf.eh.doe.gov/data/ihp1d/400199e.pdf (accessed April 1, 2008); Robert A. Conard, letter to Dr. Charles L. Dunham, Director of Biology and Medicine, June 5 1958. Subject: Recounting of Specific Things That Happened in the Marshall Islands, http://worf.eh.doe.gov/data/ihp1d/400212e.pdf (accessed April 1, 2008).

[87] See T. M. Beasley, E. E. Held, and R. M. Conard, *Iron-55 in Rongelap People, Fish and Soils* (Seattle: Laboratory of Radiation Ecology, College of Fisheries, University of Washington; Upton, NY: Brookhaven National Laboratory, Medical Department, September 1970).

[88] John H. Morrison, *Deseret Test Center Final Report on DTC Test 50-68,* cited in Ed Regis, *The Biology of Doom: The History of America's Secret Germ Warfare Project* (New York: Henry Holt and Company, 1999), 204–06.

1967, 1968, and 1969 reports in the summary of health conditions that a "rather serious outbreak of Hong Kong influenza occurred among the Rongelap people in 1968 and may have been responsible for the deaths of a 58-year old exposed woman and of an unexposed boy who died of meningitis complicating the influenza."[89]

SUMMARY OF CONSEQUENTIAL DAMAGES

Research plans and priorities established in 1954 by the U.S. government emphasized monitoring and human subject research rather than treatment of the Rongelap people. These plans shaped the nature of medical attention in subsequent years for the Rongelapese. Medical recommendations to avoid future exposures were offset by scientific research priorities. Human subject research included a long-term experiment involving the purposeful exposure of an entire population to hazards contained in a contaminated setting. Decisions to repatriate the people of Rongelap were made with full knowledge that environmental hazards were present and would pose some threat to an already exposed population, and with full intent to exploit the research opportunities associated with continued exposures. The people of Rongelap were returned to a contaminated setting, where they remained under a policy of no reevacuation, and were thus subject to further radioactive and toxic assaults from future weapons tests. The consequential damages of these policies and priorities include decades of medical focus on research questions rather than individual health care needs. These actions constitute systematic and long-term abuse of the Rongelapese—abuse that continues to generate hardships for individuals, families, the broader communities of Rongelapese, and the nation.

Difficulties of Life in a Contaminated Setting

Dorothy Amos: *Then we went back to Rongelap. They [the U.S. government] said they had cleaned the island.*[90]

---

[89] Robert A. Conard et al., *Medical Survey of the People of Rongelap and Utirik Islands Thirteen, Fourteen, and Fifteen Years After Exposure to Fallout Radiation (March 1967, March 1968, and March 1969)* (Upton, NY: Brookhaven National Laboratory, 1969), 3, http://worf.eh.doe.gov/data/ihp2/0991_.pdf (accessed October 25, 2007).

[90] Dorothy Amos, interview by Holly M. Barker, Ebeye, March 18, 1999.

When the people returned to Rongelap, they found that numerous possessions that they had not gathered and brought with them during their evacuation were missing:

Almira Matayoshi: *When they came to evacuate us after the [Bravo] bomb, they told us not to take anything at all—none of our possessions. So I left everything behind. I left the* peba in kallimur *[papers with promises that function as contracts] signed by my grandparents that showed I had land. When we returned after a few years, everything in our houses was gone. Now I can't go to court and prove I had land, because the papers are gone. They told us not to bring anything.*

When we returned, the papers and everything was gone. Everything was burned—papers, clothes, pictures. Our boats were gone. Our tools for working, including the drekeinin *that is passed down from our mothers. We don't know how to make those tools, or how to make the boats that could sail ten people at once. We had about five of those large sailing boats, but they were all gone when we returned.*[91]

The consequences of these losses were profound. In the case of Almira Matayoshi, who did not have time to collect her landowner papers, she and her offspring lost the claim to their land, because they do not have the documents necessary to prove ownership.

In addition to the difficulties of adjusting to the loss of personal, household, and family property, the Rongelapese confronted significant difficulties in adjusting to their contaminated atoll. Upon their return, they found an altered landscape, with "atypical plants, including trees with three to five crowns."[92] Rongelap trees and plants were described by University of Washington scientists in July 1957 as "mutants" because of their extra flowers and limbs and their stem abnormalities—atrophied, or "thickened, swollen" stems covered with cancerous warts.[93]

From the onset, the people of Rongelap began to experience a wider variety of health problems than previously known. Jorulej Jitiam, the health assistant posted on Rongelap in 1957, expressed his concern in a letter to the district director of public health in Majuro. He asked for treatment advice for

---

[91] Almira Matayoshi, interview by Holly M. Barker, Honolulu, June 14, 2001.

[92] T. A. Davis, *Freaks of Nature: Branching in Coconut Palms* (Kayangulam, Kerala, India: Central Coconut Research Station, 1963).

[93] Edward E. Held, letter to D. H. Nucker, High Commissioner, Trust Territory of the Pacific Islands, August 7, 1959. Subject: Regarding a Biological Survey of Rongelap, http://worf.eh.doe. gov/data/ihp1d/400222e.pdf (accessed April 1, 2008). This report concludes: Some plants "survived the bombing unharmed, but when they grew into mature plants, they soon succumbed. It is clearly indicated that they were suffering from persisting effects of the bombing. Continuing radiation from soil particles must be the cause"; Lauren R. Donaldson, *Radiologic Survey of Bikini, Eniwetok and Likiep Atolls—July–August 1949* (Seattle: University of Washington, July 12, 1950), 43, http://worf.eh.doe.gov/data/ihp1d/149062e.pdf (accessed December 17, 2007).

*impetigo that doesn't heal . . . and wounds that won't heal for up to 3 months and get worse.
. . . Such things didn't occur to these people while they were in Ejit, but now many of them
have gotten these diseases. . . . What I think is that there is still radioactivity in the island
because even if a person got a small laceration on his leg, etc., it will become infected and get
bigger. Frequent abdominal pain is being noticed among the people. . . . I want a doctor to
come here before the AEC's visit so he can have a fairly good idea of what cause these things
to be happened. Please come as soon as possible.*[94]

Although radionuclides and their bioaccumulation in foodstuffs were an invisible
threat to the Rongelapese, the people suspected that their problems stemmed from
high levels of radiation in their food. The Rongelapese got blisters in their mouths
and food poisoning from eating certain foods. The foods that made them ill after
resettlement never made them sick before the Bravo test. Because of the blisters
in their mouths from the food, and because they could see the effects of radiation
in the trees and plants they ate, the Rongelapese attributed their health problems
to the presence of radiation in their environment:

John Anjain: *We had no problem with our food before the testing. Afterwards, arrowroot
and the plants gave us blisters in our mouths. I also think the problems with our throats
[thyroids] are from the poison. After we went back, new fish gave us fish poisoning, such
as paan [red snapper]. Rongelap's lagoon was affected. I got fish poisoning twice after
returning. . . . As for the crabs, DOE told us not to eat them. It's forbidden to eat them.
Before the bomb, we used to eat them a lot. DOE were the ones that said it was forbidden
to eat them. I asked why. They told me to ask God.*[95]

Chiyoko Tamayose: *We ate foods that made our throats swell and close up, and even
made us shake like we had epilepsy. I remember this after eating crab, I don't know if it
became poisonous like some species of fish, or what. . . .*[96] *The people got mouth blisters from
eating arrowroot, fish, clams, and coconuts. . . . We had no choice but to go back to Rongelap
[in 1957]. . . . It is impossible to limit people's consumption once they are on their land.*[97]

Nerja Joseph: *Some coconuts had two or three heads. The arrowroot had no contents . . .
it was empty inside.*[98]

Dorothy Amos: *After returning, many things were different. The food we ate . . . there
was a food we didn't eat, coconut crab. The fish—our mouths were hot . . . we got burns.
Arrowroot also gave us blisters. We saw that many things were different. We sucked on*

[94] Jorulej Jitiam, letter to the district director of public health in Majuro, November 27, 1957.
On file at the Republic of the Marshall Islands Embassy, Washington, DC.
[95] John Anjain, interview by Holly M. Barker, Ebeye, March 16, 1999.
[96] Interview by Holly M. Barker, Honolulu, June 13, 2001.
[97] Chiyoko Tamayose, advisory committee discussion, March 2, 1999.
[98] Nerja Joseph, interview by Holly M. Barker, Majuro, March 8, 1999.

pandanus and that was all right, but the other foods were bad. We saw some coconut trees that had two heads, some three. We saw they weren't good for us to eat. We understood there were differences, because we grew up there and we knew. They gave us other food to help us for awhile on Rongelap. Then we ate leaves and animals.[99]

Lijon Eknilang: *The throats [thyroids] of the birds became abnormally large and swollen in the years after the bombs. When we opened them up, they had hard, white rocks in their throats that we had never seen before.*[100]

The resettled Rongelapese ate vegetables, fruits, and tubers grown in heavily contaminated soil. They ate coconut crabs, chickens, and pigs that had in turn consumed radioactive foods and had thus accumulated high concentrations of radioisotopes. Radiation also affected marine sources of food, such as fish that consumed radioactive algae and plankton. Consuming foods that accumulated and thus intensified radiation from the environment compounded the exposure of the Rongelapese. Ethnographic interviews suggest that fish poisoning, while not unknown prior to the nuclear tests, became an immediate and serious problem, with some species and locales that were previously safe causing poisoning:

People got fish poisoning from types of fish that never caused poisoning in the past, such as iol *(mullet), and* malok. *Before the tests, only the* jujukop *(barracuda) fish caused fish poisoning.*[101]

Responsibility for the long-term monitoring of the Rongelapese (1954 to 1997) was given to a U.S. government weapons facility, Brookhaven National Laboratory, with Dr. Robert Conard appointed director in 1956.[102] BNL scientists embraced their duties and research opportunities, noting that on Rongelap Island, "the levels of activity are higher than those found in any other inhabited location in the world. The habitation of those people on the island will afford most valuable ecological radiation data on human beings."[103]

---

[99] Dorothy Amos, interview by Holly M. Barker, Ebeye, March 18, 1999.
[100] Lijon Eknilang, advisory committee discussion, Majuro, March 2, 1999.
[101] Lijon Eknilang, advisory committee discussion, Majuro, March 2, 1999.
[102] A. C. Deines, D. R. Goldman, R. R. Harris, and L. J. Kells, *Marshall Islands Chronology, 1944–1990* (Rockville, MD: History Associates Incorporated, 1991), http://worf.eh.doe.gov/data/ihp2/0386_.pdf (accessed December 17, 2007). This responsibility was interpreted as prioritizing medical research over treatment. The exposed population enrolled in Project 4.1 repeatedly complained that they were given too many examinations for the sake of research but not being treated for their health problems. In response to complaints such as these, U.S. government medical researchers proposed that "perhaps next trip we should consider giving more treatment or even placebos" (see note 174). Robert A. Conard, letter to Dr. Charles L. Dunham, Director of Biology and Medicine, June 5 1958. Subject: Recounting of Specific Things That Happened in the Marshall Islands, http://worf.eh.doe.gov/data/ihp1d/400212e.pdf (accessed April 1, 2008).
[103] Robert A. Conard, Leo M. Meyer, J. Edward Rall, Austin Lowery, Sven A. Bach, Branford Cannon, Edwin L. Carter, Maynard Eicher, and Hyman Hechter, *March 1957 Medical Survey of*

Plant observations provided clear indications that radioisotopes were present in the food chain. Samples of the coconut crab—a main source of food for the people of Rongelap—were not analyzed and reported upon until February 1958, however, and it was not until June1958 that the Rongelapese were informed by the medical survey team that eating coconut crabs presented a health risk from high levels of Sr-90.[104] As the ethnographic data indicate, the June 1958 ban on coconut crab angered and confused the Rongelapese. Many people did not adhere to the restriction:

Nerja Joseph: *First they said we could eat crabs, and then they said to stop. What's the point when we already ate them?*[105]

Ken Kedi: *Culturally, you can't tell people not to eat food, like crab, they see. Even if we were hungry, they said not to eat the crabs, but we ate them anyway.*[106]

Timako Kolnij: *Even though it was bad, I ate them. I didn't think about it. The food we ate gave us blisters in our mouths.*[107]

Jerkan Jenwor: *Rongelap was an atoll with so much food—coconut, pandanus, and breadfruit. Now there is none. Because of the poison, it disappeared. When we returned, we ate the arrowroot. It gave us blisters in our mouths, but we had to eat it, especially during times of hunger. The local doctor eventually stopped us from eating it. We also ate crabs, even though we weren't supposed to.*[108]

Although people on Rongelap were eventually told not to eat coconut crabs, they were not told to avoid other terrestrial and marine food sources known to have high levels of radiation, such as arrowroot. Despite the extremely high levels of radiation in coconut crabs, researchers considered them only "intermediate in their general level of radioactivity" compared to giant clams.[109] The U.S. Navy

Rongelap and Utirik People Three Years after Exposure to Radioactive Fallout (Upton, NY: Brookhaven National Laboratory, 1958), 22.

[104] Robert A. Conard, letter to Dr. Charles L. Dunham, Director of Biology and Medicine, June 5, 1958, http://worf.eh.doe.gov/data/ihp1d/400212e.pdf (accessed April 1, 2008). Conard writes: "I must confess I didn't know there was any difference between [a land crab and a coconut crab]. When I told them they could not eat the coconut crab, they were a little peeved since they had been told by the weather station people that they could eat them."

[105] Nerja Joseph, interview by Holly M. Barker, Majuro, March 8, 1999.

[106] Ken Kedi, interview by Holly M. Barker, Barbara Rose Johnston, and Stuart Kirsch, Majuro, March 1, 1999.

[107] Timako Kolnij, interview by Holly M. Barker, Ebeye, March 18, 1999.

[108] Jerkan Kenwar, interview by Holly M. Barker, Ebeye, March 17, 1999.

[109] Staff of the Applied Fisheries Laboratory, *Radiobiological Resurvey of Rongelap and Ailinginae Atolls.*

confirmed that clams concentrated high levels of cobalt-60 (Co-60), yet the Rongelapese were never told to avoid eating clams.[110] Researchers also reported high levels of contamination in pandanus, a staple Rongelapese food. "Pandanus fruit and coconut crabs were the two food items of major concern with respect to Sr-90 content at the time of the return of the Rongelapese,"[111] yet pandanus was not restricted from their diet.

The radiation that worked its way through the plant and animal food chain that the Rongelapese depended on for survival provided scientists with the research opportunity to significantly expand their understanding of bioaccumulation and environmental sources of exposure in human beings. Thus in "the long range radiation ecology study at Rongelap major emphasis was placed on studies of the soil-plant relationship, aquatic bird populations, and mineral transport, as well as evaluations of the uptake of specific isotopes by plants and animals used as food by the natives."[112]

While it was assumed that radioactivity would decrease over time, ecological surveys by the University of Washington found persistent radiation in the soil and in terrestrial and marine organisms on Ailinginae, Rongerik, and Rongelap.[113] A 1977 study documented and measured the presence of six fallout radionuclides—Co-60, Cs-137, Eu-155, Am-241, Pu-240, and Sr-90—in soil samples taken from Ailinginae and found that Cs-137 and Sr-90 were the most abundant radionuclides.[114] On

---

[110] Weis et al., *Residual Contamination,* 19.

[111] Edward E. Held, letter to D. H. Nucker, High Commissioner, Trust Territory of the Pacific Islands, August 7, 1959. Subject: Regarding a Biological Survey of Rongelap, http://worf.eh.doe.gov/data/ihp1d/400222e.pdf (accessed April 1, 2008).

[112] Joint Task Force Seven, *Operation Hardtack: April–October 1958* (Joint Task Force Seven, August 1958), 58, http://worf.eh.doe.gov/data/ihp2a/1030_a.pdf (accessed April 1, 2008).

[113] Excretion analysis of samples from the Rongelap population found an average body burden of cesium-137 for 1954 that was twenty times the rate reported for U.S residents, declining in 1957 to a rate that was equal to the U.S. rate. In 1957, residents who had returned to Utrik in 1954 presented rates some forty-eight times greater than those of the Rongelap population. When Rongelap residents were returned to their home islands in 1958, nine months later, samples were obtained; findings demonstrated significant increases of cesium-137. A joint army–AEC study reported that "expected increases in the trace amounts of radionuclides in the food supply of a large population would afford an opportunity to invesitagte the rate of equilibrium and the discrimination factors operating between food supply and man." It concluded: "Since resettlement of the Marshallese on Rongelap atoll in July 1957, the urinary excretion level of cesium-137 has increased about 140 fold and about 20 fold for strontium. Zinc-65 was readily detected from the March 1958 survey." H. A. Woodward, A. G Schrodt, J. E. Anderson, H. A. Claypool, J. B. Hartgering. "Determination of Internally Deposited Radioactive Isotopes in the Marshallese People by Excretion Analysis," in *Fallout from Nuclear Weapons* (Washington, DC: Walter Reed Army Institute of Research and the Atomic Energy Commission, n.d.), 1330–47, http://worf.eh.doe.gov/data/ihp2/1047_.pdf (accessed April 1, 2008).

[114] V. A. Nelson, *Radiological Survey of Plants, Animals and Soil at Christmas Islands and Seven Atolls in the Marshall Islands* (Seattle: University of Washington, 1977), 9–10, http://worf.eh.doe.gov/data/ihp1c/7843_.pdf (accessed April 1, 2008).

Rongerik, seven fallout radionuclides were commonly found in surface soil samples, and "pandanus and coconut samples contained fallout radio nuclides Cs-137 and Sr-90. The edible fruit of the Pandanus had the highest Cs-137 values."[115] On Rongelap, five gamma-emitting radionuclides were measured in fish, plant, and soil samples, and "significant fallout radionuclides" were found in clams.[116] This study concluded:

> Sr-90 and Cs-137 are dominant in the terrestrial environment and in addition, Am-241 and 239, Pu-240 are also important in the soil from Rongelap atoll, Cobalt 60 and 55 Fe are predominant in the marine environment.[117]

Bioaccumulation of radioisotopes not only occurred in terrestrial realms but also in marine ecosystems. Surveys of the plants, animals, soil, and water in 1955 indicated that some of the "highest concentrations of internally deposited activity were found in marine specimens taken from the northern Rongelap lagoon."[118] Based on oceanographic surveys and water sampling after the Castle series, "one conclusion evident from these [surveys] is that total doses of 250R or more could have been accumulated throughout an area of about 5,000 square miles."[119] The movement of fish and marine foods throughout the atoll and a 5,000-square-mile area of marine exposure meant that the Rongelapese consumed radiation from their marine environment.

U.S. researchers monitored open-sea marine plankton and its role in transporting fallout in the marine food chain.[120] Researchers observed that plankton was the most sensitive indicator of radioactivity in the sea, and radiation readings in plankton were considered "representative of that available to marine food chains."[121]

---

[115] Ibid., 10–11.

[116] Ibid., 12.

[117] Ibid., 27.

[118] W. R. Rinehart, S. H. Cohn, J. A. Seiler, W. H. Shipman, J. K. Gong, *Residual Contamination of Plants, Animals, Soil, and Water of the Marshall Islands One Year Following Operation Castle Fallout* (San Francisco: National Radiological Defense Laboratory, 1955), 11, http://worf.eh.doe.gov/data/ihp1c/0865_a.pdf (accessed April 1, 2008).

[119] T. R. Folson and L. B. Werner, *Operation Castle, Project 2.7, Distribution of Radioactive Fallout by Survey and Analysis of Sea Water, March–May 1954* (Albuquerque: Armed Forces Special Weapon Project, April 14, 1959), 53.

[120] Ibid., 17.

[121] Allyn H. Seymour, Edward E. Held, Frank G. Lowman, John R. Donaldson, and Dorothy J. South. *Survey of Radioactivity in the Sea and in Pelagic Marine Life West of the Marshall Islands, September 1–20, 1956* (Seattle: University of Washington, 1957), 55; Allyn H. Seymour, *Fish and Radioactivity* (Seattle: University of Washington Laboratory of Radiation Biology, January 21, 1960), 64, http://worf.eh.doe.gov/data/ihp1c/0383_a.pdf (accessed April 1, 2008).

In a 3,300-mile survey area in the Pacific Ocean, "radioactive materials were found in the plankton samples from every station."[122]

By 1958 university researchers had discovered that fish might be concentrating radioactivity by as much as "a thousand fold" because of the radioactive plankton they consume, similar to the way in which radiation bioaccumulates in coconut crabs.[123] Researchers also observed that "the lagoon would tend to hold radiation within its system of circulation"[124] and that radiation would concentrate in the lower levels of the lagoon, where fish, such as the sturgeon fish—a species routinely consumed by the Rongelapese—would concentrate high levels of cesium.[125]

According to a 1975 assessment of the environmental monitoring data, "the most important sources of exposure to people living on Rongelap . . . are from internal deposition of radioisotopes from certain elements in the human diet, and from long term occupancy of islands having external radiation dose rates higher than natural background."[126] As plants uptake radiation from the environment and animals consume plants and smaller animals, radiation concentrates and moves through the food chain. Radiation ecology studies at Rongelap found that the highest concentrations of radiation were found in herbivore and omnivore species of fish, such as parrot fish,[127] which was frequently consumed by the Rongelapese.

Increases in gross beta radioactivity in fish were measured on Rongelap between 1954 and March 1958.[128] As late as 1962, "the highest levels of gross beta radioactivity were found in samples of algae, fish liver and muscles, and sea cucumber muscle" at Rongerik.[129] Fish is one of the most critical food sources and the main source of protein for the Rongelapese. A 1982 study found that with the exception of

---

[122] T. R. Folsom, F. D. Jennings, M. W. Johnson, *Operation Castle, Project 2.7A, Report to the Scientific Director, Radioactivity on Open-Sea Plankton Samples* (LaJolla: University of California Scripps Institution of Oceanography, n.d.), 17; Seymour et al., *Survey of Radioactivity,* 9.

[123] Seymour, *Fish and Radioactivity,* 59.

[124] *Operation Redwing: Project 2.62a, Fallout Studies by Oceanographic Methods, Pacific Proving Grounds, May–July, 1956* (Albuquerque: Sandia Base, February 6, 1961), 85.

[125] Ibid., 1, 85.

[126] N. A. Greenhouse and T. F. McCraw. *Marshall Islands Radiological Follow Up* (Upton, NY: Brookhaven National Laboratory, August 1, 1975), 2, http://worf.eh.doe.gov/data/ihp1a/0254_.pdf (accessed April 1, 2008).

[127] Lauren R. Donaldson, *Radiobiological Survey of Bikini, Eniwetok, and Likiep Atolls, July–August 1949* (Seattle: University of Washington, July 12, 1950), 145.

[128] Ralph R. Palumbo, *Gross Beta Radioactivity of the Algae at Eniwetok Atoll, 1954– 1956* (Seattle: Laboratory of Radiation Biology, University of Washington, 1959); Arthur Welander, *Radiobiological Studies of the Fish Collected at Rongelap and Ailinginae Atoll, July 1957* (Seattle: Laboratory of Radiation Biology, University of Washington, March 5, 1958), http://worf.eh.doe.gov/data/ ihp1c/7850_.pdf (accessed April 1, 2008).

[129] Ralph F. Palumbo, *Radioactivity in the Biota at Islands of the Central Pacific, 1954–1958* (Seattle: Laboratory of Radiation Biology, University of Washington, 1962), 11, http://worf.eh.doe.gov/ data/ihp1c/0430_a.pdf (accessed April 1, 2008).

coconut, fish was the primary food consumed by the Rongelapese, because theirs is an atoll area with excellent fishing.[130]

From radiation levels monitored in bird populations, U.S. government researchers concluded that the fishing area in southern Rongelap where the people were resettled had higher radiation concentrations than the restricted northern islands. Researchers also found high concentrations of radiation in birds, a terrestrial food source consumed by the Rongelapese. The birds from southern Rongelap had higher levels of radiation than birds from the north of Rongelap.[131] According to researchers, this "unexpected" finding of "higher levels of radioactivity in the tissues of the southern birds suggest the availability of a supply of food fish with a higher average radioactive content in the southern area compared with that of northern Rongelap."[132] Unfortunately, fish and birds were staples of the Rongelapese people. In the mid-1990s, the RMI Nationwide Radiological Survey tested thousands of soil, plant, and marine samples collected throughout the nation and confirmed the existence of unsafe levels of radiation at dozens of islands.[133]

The Rongelapese observed that many species of fish that did not cause fish poisoning before the nuclear tests became poisonous afterward, and some species were poisonous in some locations but not others. Some scientists have suggested a relationship between fish poisoning and nuclear testing, with damaged reefs supporting abnormally high numbers of the plankton *Gambierdiscus toxicus,* a dinoflagellate that produces ciguatera toxin. Fish feeding on the reefs absorb this plankton; ciguatera toxin accumulates in the fish, which in turn are eaten by larger fish, which concentrate the ciguatera toxin in their flesh. Humans who eat these fish suffer from vomiting, diarrhea, loss of balance, and, rarely, death. The Marshall

---

[130] W. L. Robison M. E. Mount, W. A. Phillips, C. A. Conrado, M. L. Stuart, C. E. Stoker, *The Northern Marshall Islands Radiological Survey: Terrestrial Food Chain and Total Doses* (Livermore, CA: Lawrence Livermore Laboratory, 1982).

[131] Staff of the Applied Fisheries Laboratory, *Radiobiological Resurvey of Rongelap and Ailinginae Atolls.*

[132] Ibid., 43.

[133] S. L. Simon, and J. C. Graham, "Findings of the First Comprehensive Radiological Monitoring Program of the Republic of the Marshall Islands," *Health Physics* 73, no. 1 (1997): 66–85. In this nationwide study, emphasis was placed on measuring cesium-137 in soil and plants, and in some cases americium-241 and plutonium-239 in soil. Reported levels of cesium-137 in coconut crabs were based on assumptions of bioaccumulation rates rather than actual analysis from samples. Simon and Graham did not analyze fish samples in this study. Interpretations of dose-rate levels were based on a reported natural background rate from eating fresh marine foods that has since been revised. Total background dose was estimated at 2.4 mSv y-1 and is now estimated at 1.5 mSv y-1 (William L. Robison, scientific director, Marshall Islands Program, Lawrence Livermore National Laboratory, cover letter to the RMI Nuclear Claims Tribunal, accompanying a booklet on "the radiological situation at the northern Marshall Island Atolls," October 21, 1998). The revised background dose suggests that a greater proportion of polonium-210 and lead-210 found in marine foods can be attributed to nuclear weapons testing.

Islands and French Polynesia (where the French government tests nuclear weapons) have the highest incidence of fish poisoning in the Pacific.[134]

Reports of increased incidence of fish poisoning among the Rongelapese were dismissed by U.S. researchers, who argued that fish poisoning had been known to occur before the nuclear tests. Despite their knowledge of high levels of radioactivity in locally caught fish, U.S. government scientists consistently dismissed the possibility of a link between rising incidents of fish poisoning and radioisotopes in the marine food chain. Similarly, medical survey scientists trivialized reports that the Marshallese experienced blisters in their mouth when they ate arrowroot. The scientists explained that the blisters occurred because the people did not know how to prepare arrowroot correctly.[135] Yet the Marshallese people had been preparing and eating local foods for centuries without experiencing blisters or sickness.

Ezra Riklon (former medical officer for the Trust Territory of the Pacific Islands): I know that during these times [after resettlement] the common complaint—because I was the one who translate—after they eat arrowroot, they always developed burning sensation of the throat, and constriction of the throat which caused them to have difficulty in breathing . . . and then some of them develop rash, and nausea and vomiting . . .

After eating them [arrowroot] they complain to Dr. Conard [BNL]. And he said: "Oh, this is not unusual." Even—because at that time, all over the Marshall Islands there was fish poisoning—every atoll mostly. And these are the symptoms: nausea, vomiting, [inaudible]—this is normal. So, it was, I believe it was Dr. Conard that said this may be due to fish poisoning. But these people say: "I never eat fish, I eat arrowroot. And after that, I develop this. And every time I eat arrowroot, we always get sick." And he [Dr. Conard] said: "Every people who eat arrowroot, they will all get sick." Because he said: "Oh, maybe this is allergic to food or whatever you eat." We said, "Well, if it was allergy, why every people who eat arrowroot always get sick? If it was around the same thing, why don't some people [get sick] . . . but not all people? Not all people are allergic to some kind of food, but not every people." But we—every time these people eat arrowroot, they always get sick I noticed.[136]

As carefully documented by government scientists, the people of Rongelap were consuming contaminated foods. Ethnographic evidence suggests that men, women,

---

[134] T. A. Ruff, "Ciguatera in the Pacific: A Link with Military Activities," *Lancet* 1, no. 8631 (1989): 201–04.

[135] See Robert A. Conard, *Medical Status of the Rongelap People Five Years after Exposure to Fallout Radiation* (Upton, NY: Brookhaven National Laboratory, Atomic Energy Commission, June 1, 1959), http://worf.eh.doe.gov/data/ihp1a/3200_.pdf (accessed April 1, 2008); Robert Conard, letter to Courts Oulahan, deputy general counsel, Atomic Energy Comission, April 12, 1961, http://worf.eh.doe.gov/data/ihp1b/3813_.pdf (accessed April 1, 2008); Robert A. Conard, letter to Jonathon B. Bingham, U.S. representative to the Trusteeship Council of the United Nations, May 1, 1961, http://worf.eh.doe.gov/data/ihp1b/3811_.pdf (accessed April 1, 2008).

[136] Ezra Riklon, interview by Holly M. Barker, Majuro, August 18, 1994.

children, and the elderly were exposed differently because they ate more or less of certain local foods. For example, babies and the elderly tended to eat softer foods, such as boiled pandanus. Women tended to suck on the bones and eat the organs of fish, while men ate more of the flesh. Surveys of radiation levels in fish found high levels of radioactivity in the liver and viscera of fish—the parts often consumed by women—which could lead to a greater consumption of organ- and bone-seeking radionuclides in women.[137]

Exposure to contaminated foods was widespread, as food from Rongelap was distributed throughout the Marshalls. The Marshallese depended on their neighbors and families to exchange food. Implicit in this notion of exchange was the need for people to work together to prepare food. Most frequently, people cultivated one type of food and traded it for other types. Because food preparation in the Marshall Islands was labor intensive, people benefited from an economy of time by exchanging food sources. The presence of food resources in Marshallese communities expressed the well-being of the communities. For this reason, the sharing of food was an important component of group health and keeping the community together.[138] There was an obligation for workers to share their food with their leaders at least once a year, usually at breadfruit harvesting time. Because food resources on Rongelap Island alone were not adequate to provide for the people residing on the island, or for customary distribution, the Rongelapese used the small islands in the north to collect and distribute food.[139] The northern islands of Rongelap are approximately ten times as contaminated as the southern islands, and Cs-137 rates in coconut, pandanus and breadfruit from the northern islands can increase exposure by a factor of eight to thirty-two.[140] Thus exchange of food resources from Rongelap led to the exposure of family members and iroij on distant islands:

Mike Kabua: *Contaminated foods from Rongelap were shared with [people on] Ebeye and Majuro. You can't ignore your relatives in the urban areas, because we have to share all food . . .*

Iroij visits to Rongelap after 1957 were difficult. The people were required to give food to their iroij, even when the food was contaminated, such as coconut crabs, because of the cultural importance. People didn't want to give contaminated food to the iroij, but they had

---

[137] Lauren R. Donaldson, *Radiologic Survey of Bikini, Eniwetok and Likiep Atolls—July–August 1949* (Seattle: University of Washington, July 12, 1950), 8, http://worf.eh.doe.gov/data/ihp1d/149062e.pdf (accessed December 17, 2007).
[138] Nancy Pollock, *These Roots Remain: Food Habits in Islands of the Central and Eastern Pacific since Western Contact* (Laie, Hawaii: Institute for Polynesian Studies, 1992), 195.
[139] Advisory committee discussion, March 3, 1999.
[140] National Research Council, *Radiological Assessments for Resettlement of Rongelap in the Republic of the Marshall Islands* (Washington, DC: National Academy Press, 1994), 83.

to. By the same token, the iroij didn't want to accept the contaminated food, but had to. The iroij was scared of eating coconut crab, but it is so important to give coconut crab to the iroij that the iroij could kick people off the land for not giving it to him. The people were also afraid the iroij would reject the food.[141]

DOE studies also documented the presence of radiation in the well and cistern water sources that the Rongelapese depended on daily for survival. In 1957, the year the U.S. government resettled the Rongelapese, researchers determined that the contamination level in Rongelap's water was higher than had been reported in earlier surveys.[142]

In addition to food and water sources, inhalation is another major pathway for radiation exposure.[143] The Rongelapese inhaled radioactive fallout from the Bravo test in 1954. After resettlement, the Rongelapese were vulnerable to the inhalation of radioactive dust resuspended in the air after ground-moving activities, with different activities by men and women contributing to different levels of exposure. For example, women often used coconut husks to fuel cooking fires, while men burned palm fronds and other vegetation in cultivation areas. In April 1999, the Rongelap community revisited their home island to find a very dusty, dry island. Walking around Rongelap Island in conjunction with this project, researchers and informants were literally covered with black dust. People pulled their shirts over their mouths for fear of breathing contaminated soil. People also noted extreme changes in the vegetation and expressed concern that observed changes were the long-term consequence of radioactive contamination. They were surprised and dismayed to see yellow, dry, and barren plants and trees. Few fruits were found on the coconut, pandanus, and other food-producing trees.

Radiation deposited in the soil from the weapons tests leads to both internal and external exposure in human beings: "Soil contamination provided the basis for human exposure in two ways. Radiation that emanated from the ground or standing vegetation led to external dose. Radiation that emanated from food and water into the human body were responsible for internal dose."[144] In addition to internal and external pathways to human exposure to radiation in the soil, "intake through the skin can be either through the intact skin or through wounds. Of particular interest in the Marshall Islands resettlement is the potential for uptake from contaminated soil."[145] Radiation scientists were especially concerned about children and exposure to radiation from soil:

---

[141] Mike Kabua, advisory committee meeting, Majuro, March 2, 1999.

[142] Lauren R. Donaldson, University of Washington, Applied Fisheries Laboratory, letter to A. H. Seymour, January 11, 1957. Subject: Difference in activity levels in the water at Rongelap as reported in earlier visits compared with a survey done in July 1956, http://worf.eh.doe.gov/data/ihp1d/400199e.pdf (accessed April 1, 2008).

[143] National Research Council, *Radiological Assessments,* 68.

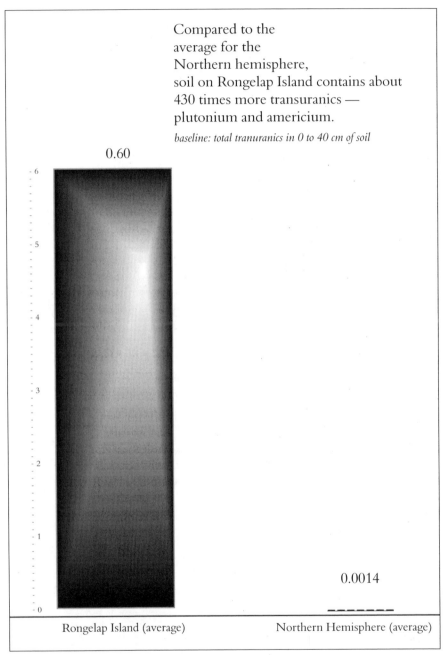

Compared to the average for the Northern hemisphere, soil on Rongelap Island contains about 430 times more transuranics — plutonium and americium.

baseline: total tranuranics in 0 to 40 cm of soil

0.60

0.0014

Rongelap Island (average)   Northern Hemisphere (average)

Figure 1. Levels of plutonium-239/240 and americium-241
Source: Bernard Franke, citing findings from 1978 restudy of radiological conditions on Rongelap. Testimony to the U.S. House Committee on Interior and Insular Affairs, Subcommittee on Insular and International Affairs (November 16, 1989).

One of the key exposure pathways for plutonium may well be infants eating contaminated soil. All over the world, infants crawl on the ground and eat dirt. Since the plutonium on Rongelap does not seem to be equally distributed, an infant could ingest a small amount of soil with one bite containing what one could call a "hot spot." Another exposure pathway could be direct uptake of plutonium into the bloodstream with wounds.[146]

Human ecological studies became an important component of U.S. government research on radiation effects in Rongelap. Clear linkages were established between levels of radiation found in plants and levels of radiation found in animals that consumed the plants. For example, the extraordinarily high levels of radiation found in the organs of rats on Bikini and Enewetak were roughly in proportion to cesium levels in plants on the various islands. Apparently, cesium-137 is concentrated in the muscle, liver, kidney, gut, skin, and bone. It was noted that the bone-seeking elements were likely to be the major cause of hazard for long-term internal effects.[147]

In the plants exposed to radiation, at a certain point "cell division ceased and the cells enlarged and took on an abnormally mature appearance."[148] Interestingly, this change corresponds with the hydatiform molar pregnancies experienced by Marshallese women after the testing. These pregnancies, called grape babies by the Marshallese, result when cells stop dividing and swell to the size of grapes.[149] The enlarged cells attach themselves to the uterus and are often miscarried several months later, giving the impression of a birth to grapes:

Almira Matayoshi: *I was pregnant when they dropped the bomb [Bravo]. I was flown off of Rongelap with the other pregnant women and the elderly people. The rest of the people left on the boat. I gave birth to Robert on Ejit, and he was normal.*

The child I had after Robert, when we had returned to Rongelap, I gave birth to something that was like grapes. I felt like I was going to die from the loss of blood. My vision was gone,

[144] Henry Kohn, *Rongelap Reassessment Project Report* (Berkeley, CA: Rongelap Reassessment Project, 1989), 19, http://worf.eh.doe.gov/data/ihp1d/400005e.pdf (accessed April 1, 2008).
[145] National Research Council, *Radiological Assessments*.
[146] Bernard Franke, testimony to the House Committee on Interior and Insular Affairs, Subcommittee on Insular and International Affairs, November 16, 1989, 9.
[147] Arthur Welander, Kelshaw Bonham, Lauren R. Donaldson, Ralph F, Palumbo, Stanley P, Gessel, Frank G. Lowman, and William B. Jackson. *Bikini-Eniwetok Studies, 1964,* part 1: *Ecological Observations* (Seattle: Laboratory of Radiation Biology, University of Washington, September 15, 1966), 45, 97, http://worf.eh.doe.gov/data/ihp1d/19902e.pdf (accessed April 1, 2008).
[148] Donaldson, *Radiologic Survey,* appendix 9.
[149] Glenn Alcalay, testimony to the White House Advisory Committee on Human Radiation Experiments regarding incidence of reproductive abnormalities and distance from ground zero, March 17, 1995.

and I was fading in and out of consciousness. They emergency evacuated me to Kwajalein, and I was sure I was going to die.

After the grapes, I had a third child. It wasn't like a child at all. It had no bones and was all skin. When I gave birth they said, "Ak ta men en?" [What is that thing?] Mama said uror [a term denoting exacerbation]. It was the first strange-looking child that people had seen. I was the first.

That time was the worst time in my life. The two times I gave birth to those things was the worst suffering of my life. I feel both angry and embarrassed. Many of the children born on Rongelap died. Now we think our children need medical attention, but they [the U.S. government] won't take care of them.[150]

While the U.S. government continued to assure the Rongelapese that living on their atoll presented no threat to their health, it was evidently concerned about the ill effects of short-term exposures to its own workers. Consider the June 27, 1977, examination and analysis of body burdens of a Peace Corps worker who spent eight months on Rongelap teaching elementary school (1975–1976), three months on Majuro, and an additional nine months on Rongelap (1976–1977):

She mentioned that she did swim in the lagoons "a bit" although this practice was generally forbidden to women. She lived in a plywood dwelling and ate rice, flour, fish, coconut meat and coconut crabs. She is 25 years old, does not smoke ("maybe occasionally") and has never had nuclear medical procedure involving the administration of any radionuclide.[151]

Urine samples and measurements from the whole-body counter found the presence of cesium-137 in the Peace Corps worker (measured body burden of 45.2 nci).

In addition to Rongelap, researchers were also aware that people were ingesting large amounts of radiation from the environments of other areas in the Marshall Islands, most notably Bikini. In a 1976 letter to James Livermen, assistant for environmental health and safety at the Energy Research and Development Administration, Dr. Robert Conard noted:

As you know, by copy of Ed Hardy's letter to me of 7-6-76, the repeated plutonium analyses on the urine samples of people living in Bikini obtained last April again showed the same levels previously reported in his letter of March 26, 1976. Disturbing also was the finding

---

[150] Almira Matayoshi, interview with Holly M. Barker, Honolulu, June 13, 2001. U.S. law specifies that the Department of Energy provide health care services only to those people who resided on Rongelap, Rongerik, and Ailinginae on March 1, 1954. Those resettled in 1957 or later, and the offspring born on those islands, are not eligible for medical services from DOE.

[151] Norman Cohen, letter to Robert Conard, August 30, 1977, http://worf.eh.doe.gov/data//ihp2/1618_.pdf (accessed October 25, 2007).

of similar low levels of plutonium in the urines of residents of Rongelap Atoll. One wonders if perhaps people living in other areas of the Marshalls, not directly involved in the fallout, might also show increased levels because of proximity to the proving grounds. We intend to check this.[152]

Despite Conard's intentions, a nationwide survey has never been accomplished, although documents disclosed in the1990s clearly indicate nationwide exposure to significant fallout.

Throughout their stay on the contaminated atoll, the Rongelapese expressed concern that they had been returned to a contaminated setting and that they were suffering from repeated incidents of fallout from additional weapons tests. In 1959 Rongelap magistrate John Anjain formally voiced his suspicions that people were allowed to consume foods contaminated by radiation and recorded his concerns in his magistrate's record book. On the occasion of a 1959 visit of a United Nations mission to Rongelap, Anjain asked:

What is the difference between coconut crabs [which became restricted in 1958] and other kinds of foods? Why are crabs bad, but not other foods? Like the crabs, pigs also eat things from the ground, so people must eat poisonous pigs![153]

Comments from an oral history interview with James Robertson, a doctor who examined the Rongelapese after their initial exposure to Bravo and in subsequent medical surveys, and who used Rongelapese subjects in procedures and experiments at Brookhaven, provide an after-the-fact answer to Anjain's questions:

We couldn't really measure the radiation exposure because the external gamma rays don't leave any residual radioactivity. So that had to be determined by more indirect methods. But what we could measure was the radioactivity that had gotten into the body from their ingesting contaminated food. We had a gamma ray spectrometer that could distinguish between the peaks, and we could tell what [peaks were signatures of what elements] and we made records of this.[154]

---

[152] Robert Conard. letter to James Livermen, assistant for environment and safety at the Energy Research and Development Administration, 1976, http://worf.eh.doe.gov/data/ihp1a/1376_.pdf (accessed December 17, 2007).
[153] John Anjain, record book, 1959, 12; Robert Conard, *An Outline of Some of the Highlights of the Medical Survey of the Marshallese Carried out in February–March 1959, 5 Years after the Fallout Accident* (Upton, NY: Brookhaven National Laboratory, 1959), 2, http://worf.eh.doe.gov/data/ihp1a/2698_.pdf (accessed April 1, 2008).
[154] James S. Robertson, oral history interview by Michael Yuffee and Prita Pillai, U.S. Department of Energy, January 20, 1995, http://hss.energy.gov/healthsafety/ohre/roadmap/histories/0478/0478toc.html (accessed December 17, 2007).

SUMMARY OF CONSEQUENTIAL DAMAGES

Living in a contaminated setting meant cumulative exposures to individuals as radiation moved from the soil to the plants and animals and through the food chain. The U.S. government monitored the movement of radionuclides concentrated in marine and terrestrial foods as they moved through the ecosystem and the bodies of the Rongelapese. The U.S. government purposefully failed to warn the Rongelapese of the presence of radiation in their food, soil, water, fish, and local environs and to restrict their consumption of local foods. As a result, the Rongelapese suffered continual discomforts from consuming contaminated food. Their health continues to be affected from the cumulative effects of these exposures.

Degenerative Health and Health Care Issues on Rongelap

From 1957, when the Rongelapese were resettled on their homeland, to 1972, AEC and its successor, the DOE, conducted annual medical exams of the Rongelapese. After the September 1972 discovery of leukemia in Lekoj Anjain, the teenage son of John Anjain, the DOE initiated twice-yearly exams for the population. Medical survey visits involved documentation of conditions, harvesting of samples for various biomedical research studies, and—after thyroid problems were identified in the 1960s—treatment for specific radiation-related ailments. As noted in the Advisory Committee on Human Radiation Experiments (ACHRE) review: "Primary case, however, remained inadequate. There were serious epidemics of poliomyelitis, influenza, chicken pox, and pertussis, all of which, according to Dr. Conard, were imported into the Marshalls by the U.S. medical teams."[155]

In some cases the serious epidemics were predicted and, if requests for vaccines had been met, preventable. Dr. James P. Nolan writes in a June, 18, 2001, letter published in the *New York Times*:

> *In the mid-1950's, as a medical officer in the Navy, I participated with a team from the Brookhaven National Laboratory in examining inhabitants of an atoll who were irradiated by fallout from a poorly planned nuclear test that left their island uninhabitable. Less well known is the occurrence of a severe outbreak of polio in the islands in the late 1950's. This outbreak occurred several years after the mass vaccination of our own population. Despite being a United Nations trust territory under United States control, the Marshall Islands natives*

---

[155] Advisory Committee on Human Radiation Experimentation, *Final Report*, 593 (see prologue, n. 20).

never received the polio vaccine. And despite my inquiries to the Department of Interior at that time, no explanation for this inexcusable health oversight was given. [156]

A review of DOE archives finds that in January 1958, several cases of polio were reported in the Marshall Islands, and an order was issued for all military personnel participating in the 1958 Operation Hardtack to be vaccinated for polio. The failure to provide vaccines for the Marshallese led to a serious epidemic of poliomyelitis type 1 in the Marshall Islands. In 1963 on Rongelap, the medical survey team found twenty-two paralytic cases in children and three cases in adults, with one death. [157]

The Rongelapese complained bitterly about the medical survey exams and procedures from the beginning, saying they felt like guinea pigs for the United States to pry and prod rather than patients with health care concerns. Marshallese complaints that their health care needs were not being treated was acknowledged in a 1958 letter by Robert Conard, who suggested the use of placebos:

I found that there was a certain feeling among the Rongelap people that we were doing too many examinations, blood tests, etc. which they do not feel necessary, particularly since we did not treat many of them. Dr. Hicking and I got the people together and explained that we had to carry out all the examinations to be certain they were healthy and only treated those we found something wrong with. I told them they should be so happy so little treatment was necessary since so few needed it . . . etc., etc. Perhaps next trip we should consider giving more treatment or even placebos. [158]

Marshallese often use songs to express difficult emotions. The callous treatment by U.S. doctors and the sense of being passed around from one doctor to another for examination was captured by the Rongelapese in the following song:

---

[156] James P. Nolan, "Sad Tales of the Pacific Islands," letter to the editor, *New York Times*, June 18, 2001.

[157] R. A. Conard, *Preliminary Statement of the Medical Surveys of the Rongelap and Utirik People, March 1963, Nine Years after Exposure to Fallout Radiation* (Upton, NY: Brookhaven National Laboratory, 1963), 1, http://worf.eh.doe.gov/data/ihp1a/2749_.pdf (accessed October 25, 2007). See the order to vaccinate U.S. military personnel visiting the Marshall Islands, Atomic Energy Commission, minutes of 1,328th meeting, January 31, 1958. Subject: Polio vaccinations for Hardtack observers, http://worf.eh.doe.gov/data/ihp1c/8860_.pdf (accessed October 25, 2007). Other incidences of immunization requests being denied were documented in 1976 by Konrad Kotrady, who notes, "The immunization clinic was limited by the fact that the Trust Territory declined a request to provide an oral supply of polio vaccine. A diabetic clinic could not be held because the Trust Territory failed to respond to a request for hypoglycemic agents." Konrad Kotrady, *June Quarterly Trip Report for 1976* (Upton, NY: Brookhaven National Laboratory, 1976), http://worf.eh.doe.gov/data/ihp2/2186_.pdf (accessed October 25, 2007).

[158] Advisory Committee on Human Radiation Experimentation, *Final Report*, 593 (see prologue, n. 20).

Rube im kalikar ialin jen Robert non LoRauut
Bunrokean ko ion tol ien wot Lomejenma
Bun-Nineaan ko ion tol lanin im raan dron
Jen na ubon im ban ke kim jo ro-koean non Laukukot
Oh Lotalim ej jutak wot
Jekron bwe Labija ej watch raan im bon

*[Show the way from Dr. Robert (Conard's) examination room to LaRaut (Mr. Urine
Collector)*
*Over the hill to the right is Lomejenma (Mr. He's So Close When He Examines Ears,
Eyes, Nose, Throat That He Can Almost Kiss His Patients)*
*Over the hill to the left is Laninimrandron (Mr. Call the Numbers or Names of People
and Escort Patients to See Doctors in the Examination Rooms)*
*From the chest to the back is examined by Laukukwot (Mr. Put the Patients on Rotating
Equipment)*
*Oh Lotalim (Dr. Touching and Examining Both Internal and External Parts) is at ease
while LaPija (Mr. x-ray Specialist) is processing film day and night.]*[159]

By 1972 the Rongelapese were so angry about the experimental nature of the
medical program that they refused to host a DOE medical survey visit that spring.[160]
Their complaints did not go unnoticed by the international community, and when
exams continued in September 1972, they included four medical observers who
reported back to the Congress of Micronesia. It was during this survey that, after
eighteen years of intensive medical research, medical survey staff made some effort
to talk to the Marshallese about exam procedures and to advise territorial health
staff about patient health needs:

*In the interest of trying to promote a better communication between examining doctors and
the Marshallese examined, an attempt was made at the completion of each examination
to explain to the person through an interpreter the general results of the examination
and possible treatment recommended. At each island clinical conferences were held by the
physicians, including the medical observers, Dr. Riklon and the health aide, to evaluate all*

[159] This song was collected and translated by Senator Abacca Anjain-Maddison.
[160] The 1972 protest is also documented in H. Kohn, *Rongelap Reassessment Project Report to the
President and Congress of the United States,* corrected ed. (Berkeley, CA: Rongelap Reassessment
Project, March 1, 1989), http://worf.eh.doe.gov/data/ihp1d/400005e.pdf (accessed April 1, 2008).
Eventually, some of the concerns expressed during the strike became the basis of a lawsuit. "In
1972 and 1973 Rongelap again began a law suit for the damage to the 'ri-bomb' [the people of
the bomb]. The result was $25,000 for the people who had their throats cut [thyroid surgery]."
John Anjain, record book, 97. Anjain also kept a record of all people who had thyroid disorders
and surgeries (102–03).

cases examined and to recommend treatment and disposition. In some cases, the health aide was advised as to further treatment.[161]

It was also during the September 1972 survey that Lekoj Anjain was recognized as exhibiting the symptoms of leukemia. Shortly thereafter, Lekoj, along with other Marshallese patients who had thyroid nodule diagnoses, was sent to Brookhaven National Laboratory in New York for further study and treatment. Acute myleogenous leukemia was diagnosed. Other Marshallese patients received treatment and were sent home. Lekoj Anjain remained in the United States to undergo chemotherapy, transfusions, and related procedures. He died on November 15, 1972, at the National Institutes of Health at Bethesda, Maryland.

A January 1995 oral history interview with Dr. James Robertson provides possible insights into the experimental nature of leukemia treatment at that time. In this interview excerpt, Robertson refers to his work with Dr. Cronkite, one of the BNL researchers responsible for the monitoring, study, and treatment of Marshallese patients:

Robertson: *Dr. Cronkite is a hematologist. I did work with Cronkite. A thing that he got me involved with was effects of radiation on the survival of red blood cells. You asked if we treated leukemia at Brookhaven, and I'd forgotten about this. A possible way of treating leukemia was by what they call extracorporeal irradiation of the blood. So what you tried to do was, you send the blood through this machine that had a radiation source in it. I designed the radiation source for his extracorporeal irradiation machine. The idea was to irradiate the blood so that you'd kill the white cells, which are very susceptible, but then you'd also damage the red cells. So to a certain extent, a study that I got involved in was a dose and effects [study] on the red cells in connection [with] extracorporeal irradiation. I don't think extracorporeal irradiation ever developed outside of the research laboratory into a treatment for leukemia, but it was an idea.*

Yuffee: *Was it tested on humans?*

Robertson: *Oh, yes.*

Yuffee: *Without much success?*

---

[161] Brian Farley, Special Joint Committee Concerning Rongelap and Utirik Atolls, Congress of Micronesia, report to Robert Conard, October 13, 1972, http://worf.eh.doe.gov/data/ihp2/3070_.pdf (accessed October 25, 2007).

Robertson: *Well, there was success of a sort, in that it would decrease the lymphocyte 105 population without too much red cell damage. But the thing is, the red cells then are being regenerated at such a rapid rate that it didn't have any permanent effect.*[162]

A *Newsweek* reporter who shared a room with Lekoj prior to his death observed that:

Lekoj spends a good deal of his time curled up in a ball on his bed, wrapped in a blanket, only his curly black hair and his brown skin showing above the blanket. He speaks very little English, but he grins broadly, showing strong white teeth, when anybody smiles at him. . . . 'How are you feeling?' I asked Lekoj this morning, as I do most mornings. 'No good. Deezy,' said Lekoj. The powerful chemicals he is taking cause an acute nausea. . . . Lekoj gets blood transfusions . . .

[W]hy should this charming young man (age 19), who should be chasing fish or possibly girls in some sparkling Pacific lagoon, suffer this lonely and outrageous fate, and by man's doing, not by God's?[163]

A February 1974 report presented to the Fifth Congress of Micronesia by the Special Joint Committee Concerning Rongelap and Utirik Atolls summarizes some of these events and concerns. It reports the findings from a study examining the psychological effects of the 1954 incident, noting,

People do worry quite a lot as a result of their experience, not only for themselves, but also for their children. . . . What is perhaps the most interesting fact is that the exposed and unexposed alike are still afraid to eat local food or live on their islands. It is as though the fear of illness from contamination persists in their minds like the residual radiation which still exists in the food chains and ecology of their islands.[164]

The Congress of Micronesia report recommended compensation for a variety of illnesses caused by radiation, citing as justification:

[162] James S. Robertson, oral history interview by Michael Yuffee and Prita Pillai, U.S. Department of Energy, January 20, 1995, http://hss.energy.gov/healthsafety/ohre/roadmap/histories/0478/0478toc.html (accessed December 17, 2007).
[163] Stewart Alsop, "The First Hydrogen Death, Another Micronesian Claim to Fame. Further Rembrances on the Incident," *Micronesian Independent,* February 22, 1974, reprinted in Borja et al., *Compensation for the People of Rongelap and Utirik by the Special Joint Committee Concerning Rongelap and Utirik Atolls to the Fifth Congress of Micronesia* (Congress of Micronesia, February 28, 1974), appendix 8, http://worf.eh.doe.gov/data/ihp2/1428_.pdf (accessed April 1, 2008).
[164] Ibid., 29–32.

Concerning the giving of radioactive materials to patients . . . the prudent assumption is that all ionizing radiation to the patient is harmful. Consequently, the Committee position that whether or not "damage" can be proven is irrelevant, since it is a fact that exposure occurred, and that since exposure to radiation is harmful, then it is highly probable that damage did indeed occur.[165]

This report, and subsequent communications between the Congress of Micronesia and the U.S. government, prompted a number of actions by medical survey staff meant to diminish public attention and concern. Later in 1974, Robert Conard attended a formal meeting on Rongelap Island. He addressed questions raised by the Rongelap Fallout Survivors Association:

I would like to say a few words about the little bit of radiation that is left on Rongelap Island. Some of the people have been worried about radiation in the food. Some will not eat arrowroot flour because they believe that it contains "radiation poison." I would like to say that with the exception of coconut crabs from the northern islands of the Rongelap Atoll, all foods can be eaten.

. . . [In response to a request to provide food at the time of examinations]: We had not been aware that so much time was involved in gathering food for the families . . . in the future, in view of this request, we shall attempt to serve prepared food to the members of the families on the day they come in for examinations.

. . . [In response to questions concerning the need for thyroid operations on Shiko, a member of the "nonexposed" group]: We do not believe the unexposed Rongelap people will get any radiation sickness from living on Rongelap since there is practically no radiation left on the island.

. . . [In response to questions about how much longer the Brookhaven team planned to continue experiments on Rongelap]: These examinations should not be called experiments. In no way have we carried out experiments on the people.[166]

---

[165] Ibid., 29.

[166] Robert Conard, Brookhaven National Laboratory, account of meeting with Rongelap People, April 1974, http://worf.eh.doe.gov/data/ihp2/2833_.pdf (accessed December 17, 2007). Typically, when the Rongelapese complained about the medical program, AEC and DOE doctors used data they had collected to make it appear that the Rongelapese were not suffering from any medical ailments with the exception of thyroid disorder. Time and time again, the U.S. government used information gathered from a faulty control group to downplay the effects of radiation: The United States called the population that was not present on Rongelap for the Bravo test, but was resettled on Rongelap in 1957 and ingested radiation from a severely contaminated environment, unexposed. When the "unexposed" group exhibited the same medical problems as the group "exposed" to radiation from Bravo, the U.S. government maintained that the "exposed" group was in good health because people in the "unexposed" group suffered similar illness rates. While minimizing the effects of radiation to the Marshallese and the world, U.S. government researchers continued to collect data from the Rongelapese.

Despite these efforts to placate fears, the Rongelap community continued to express anger and concern over their experiences as human subjects in AEC research. The sentiments of the community were captured in an April 9, 1975, letter to Dr. Robert Conard by Nelson Anjain, who was magistrate at the time:

I am writing to you to clarify some of my feelings regarding your continued use of us as research subjects.

I realize now that your entire career is based on our illness. We are far more valuable to you than you are to us. You have never really cared about us as people—only as a group of guinea pigs for your government's bomb research effort. For me and for other people on Rongelap, it is life which matters most. For you it is facts and figures. There is no question about your technical competence, but we often wonder about your humanity. We don't need you and your technological machinery. We want our life and our health. We want to be free.

In all the years you've come to our island, you've never really treated us as people. You've never sat down among us and really helped us honestly with our problems. You have told people that the "worst is over," then Lekoj died [the author's nephew]. I don't know yet how many new cases you'll find during your current trip, but I am worried that we will suffer again and again.

I'll never forget how you told a newspaper reporter that it was our fault that Lekoj died because we wouldn't let you examine us in early 1972. You seem to forget that it is your country and the people you worked for who murdered.

. . . I've made some decisions that I want you to know about. The main decision is that we do not want to see you again. We want medical care from doctors who care about us, not about collecting information for the U.S. government's war makers.

We want a doctor to live on our island permanently. We don't need medical care when it is convenient for you to visit. We want to be able to see a doctor when we want to. America has been trying to Americanize us by flying flags and using cast-off textbooks. It's about time America gave us the kind of medical care it gives its own citizens.

We've never really trusted you. So we're going to invite doctors from hospitals in Hiroshima to examine us in a caring way.

We no longer want to be under American control. As a representative of the United States, you've convinced us that Americans are out to dominate others, not help them. From now on, we will maintain our neutrality and independence from American power.

There will be some changes made. Next time you try to visit be prepared. Ever since 1972 when we first stood up to you, we've been aware of your motives. Now that we know that there are other people in this world willing to help us, we no longer want you to come to Rongelap.[167]

---

[167] Nelson Anjain, letter to Dr. Robert Conard, April 9, 1975, http://worf.eh.doe.gov/data/ihp21976_.pdf (accessed October 25, 2007).

While angry about their mistreatment in the medical survey program, the Rongelapese also realized that they had no choice but to accept continued exams, as they needed access to medication, such as the thyroid replacement medication used to keep their bodies regulated:

Rokko Langinbelik: *One of the worst things for me is that they told me to take my medication for my thyroid every day and not to forget a day, because it will shorten my life. This really upsets me, because sometimes I forget and I don't want to shorten my life.*[168]

The medical survey team continued to monitor biophysical change in the Rongelapese community and began to treat some of the emerging ailments identified as radiation related. Concurrently, beginning with samples taken immediately after exposure to fallout from the Bravo test, the U.S. government monitored cesium-137 and other isotope body burdens present in the Rongelapese —burdens that were increasing over time. U.S. government scientists did not report the increased burden to the Rongelapese.[169] Their data indicated that following resettlement in 1957, "the average adult male Rongelap body-burden rose 56% for adult men and 82% for female children."[170] In the case of strontium-90, U.S. researchers had determined that the human body absorbed roughly "70 percent of the concentration in the topsoil on which people live."[171] The first measurements of plutonium in the urine of the Marshallese were identified in 1973 (in Marshallese workers on Bikini). Because the readings were extremely high, researchers could not conclude "whether measured plutonium concentrations were excreted from the bodies or simply from some grains of the island's sand in the sample."[172] Whether the plutonium was already ingested or was still detectable in the environment, plutonium exposure was still a recorded threat to the Rongelapese.[173]

---

[168] Rokko Langinbelik, interview by Holly M. Barker, Honolulu, June 13, 2001.

[169] Jeton Anjain, testimony to the House Committee on Interior and Insular Affairs regarding the safety of Rongelap Atoll, 1989.

[170] Edward T. Lessard, letter to J. W. Thiessen. Enclosure: *Protracted Exposure to Fallout: The Rongelap and Utirik Experience* (Upton, NY: Marshall Islands Radiological Safety Program, Brookhaven National Laboratory, September 13, 1983), 12–13, http://worf.eh.doe.gov/data/ihp1c/0791_a.pdf (accessed April 1, 2008). This paper was later published by Lessard in the *Journal of Health Physics* with the conclusion that the body burden of Rongelap people exceeded allowable levels.

[171] Edward T. Lessard, *Field Trip Report* (Upton, NY: Marshall Islands Radiological Safety Program, Brookhaven National Laboratory, July 1982), 2.

[172] Franke, Committee on Interior and Insular Affairs, 7.

[173] For a summary of 1970s-era findings from whole-body counting of cesium-137 and radiochemical analysis of urine samples for strontium-90, cesium-137, and plutonium 239/240, see R. A. Conard, S. Cohn, N. Greenhouse, J. Naidu, *Summary of Radiological Monitoring of Personnel and Environment at Bikini Atoll, 1969–1976* (Upton, NY: Brookhaven National Laboratory, 1977), http://worf.eh.doe.gov/data/ihp1c/0058_.pdf (accessed April 1, 2008).

In 1989 scientists reported that the "soil on Rongelap Island contains about 430 times more transuranics (plutonium and americium) than the average transuranic levels for the Northern Hemisphere" and that, on average, "one gram of soil from Rongelap Island contains more plutonium than a North US citizen carries in his body from worldwide fallout."[174]

In 1994 the National Academy of Sciences reported the presence of five isotopes of concern to people residing on Rongelap Island: strontium-90, cesium-137, plutonium-239, plutonium-240, and americium-241. "Of these, cesium-137 account for more than 90% of the estimated [human] dose, strontium-90 provides the second most significant contribution at 2–5%."[175] Since these isotopes were of concern in 1994, it is apparent that the presence of these isotopes would have been of even greater concern in 1957 before nearly forty years of depletion.

The exposure to environmental sources of radiation in the group resettled in 1957 has affected their health. One physician found that in the DOE comparison group (people not present on Rongelap in 1954 but resettled on Rongelap in 1957), 63.6 percent of adult males and 76.8 percent of adult females had medical problems.[176] The higher incidence of female illness in the environmentally exposed is noteworthy given the startling lack of difference between incidence of medical problems in the men and women exposed to external radiation in 1954.[177]

The Nuclear Claims Tribunal currently recognizes thirty-five medical conditions presumed to be the result of the nuclear testing program, including leukemia; multiple myeloma; lymphomas; tumors of the salivary gland, parathyroid gland, and brain; cancers of the thyroid, breast, pharynx, esophagus, stomach, small intestine, pancreas, bile ducts, gallbladder, liver, colon, urinary tract, ovary, central nervous system, kidney, rectum, cecum, and bone; nonmalignant thyroid disease; hypothyroidism; growth retardation due to thyroid damage; unexplained hyperparathyroidism; unexplained bone marrow failure; meningioma; radiation sickness; beta burns; nonmelanoma skin cancer in individuals diagnosed as having suffered beta burns.

Not surprisingly, the Rongelapese were aware of the human health consequences of their internal and external radiation exposure. Many of their illnesses were outside the narrow parameters of U.S. government research, however, which tended to focus on the thyroid and the measurement of body burdens:

---

[174] Franke, Committee on Interior and Insular Affairs, 3, 6.
[175] National Research Council, *Radiological Assessments,* 1.
[176] Rosalie Bertell, statement to the House Committee on Interior and Insular Affairs, Subcommittee on Insular and International Affairs, November 16, 1989, 1, http://www.ratical.org/radiation/inetSeries/wwc3_7.txt (accessed April 1, 2008).
[177] Ibid., 10.

Catherine Jibas: *I was away from Rongelap in 1954. I returned to Rongelap . . . in 1957 and I saw friends and relatives who were afflicted with illnesses unknown to us. Their eyesight deteriorated, their bodies were covered with burnlike blisters, and their hair fell out by the handful.*[178]

Aruko Bobo: *Prior to his [Hiroshi's] death, he became insane and would wander around the island frightening everyone he met. At times he would ram his head continuously against anything near at hand and thrash around exhibiting signs of great pain. I think he finally died of the pain he felt in his head.*[179]

Nerja Joseph: *People died of radiation but we had no doctors or charts to prove that it was radiation. . . . I have scars on my hands now. No medicine will take them away, Marshallese or American. They itch and they are bothersome. My thyroid is gone. I have to take medication every day. When I take too much medication, I have too much energy —I can't sleep. I walk around and around, but I can't tire myself out to rest. When I don't take enough, I gain weight and get lethargic.*[180]

Timako Kolnij: *They found something behind my stomach. They told that to the doctor. I was operated on, and they showed it to me. It was kind of shiny. They removed a black, hard, shiny object from just under my stomach. Maybe it came from the poison. After my surgery, my health was bad.*[181]

Dorothy Amos: *It's really sad. Now we go to doctors. Many have thyroid problems. Some don't. Some have had operations twice [on their throats]. Some three times. We take medicine to help our thyroids. One pill every day. Don't forget one day—keep taking it, keep taking it. Sometimes I forget, but I'm not supposed to miss a single day. . . . Sometimes my throat really hurts, all the way down the esophagus. There are times when my head goes numb, my leg, my hand.*[182]

Because children are smaller than adults, a dose of radiation causes a larger body burden in a child than in an adult. For this reason, and because children's bodies are in the process of developing and growing, born and unborn children are at great risk from radiation exposure. Both the U.S. government and the Rongelapese maintained concerns about Rongelapese children exposed to radiation from Bravo and children who were born and raised on Rongelap Island.[183] And U.S. government

---

[178] Catherine Jiba, interview by Holly M. Barker, Majuro, August 23, 1994.

[179] Aruko Bobo, interview by Holly M. Barker, Ebeye, August 27, 1994.

[180] Nerja Joseph, interview by Holly M. Barker, Ebeye, March 8, 1999.

[181] Timako Kolnij, interview by Holly M. Barker, Ebeye, March 18, 1999.

[182] Dorothy Amos, interview by Holly M. Barker, Ebeye, March 18, 1999.

[183] See the long-term study of radiation and its effects on childhood growth, reported in W. W. Sutow, R. A. Conard, and K. M. Griffith, "Growth Status of Children Exposed to Fallout Radiation on Marshall Islands," *Pediatrics* 36, no. 5 (1965): 721–31. With regard to miscarriage and stillbirths,

reports demonstrate concern about several forms of cancer and genetic disorders in offspring.[184]

Despite concern that radiation exposure produced intergenerational effects, AEC and BNL scientists responded with denials to Marshallese complaints that their rates of miscarriage and abnormal births were linked to radiation exposure. They typically cited the work of James Neel. For example, an October 1976 letter to Congressman Balos from Dr. James Livermore concerning genetically inherited radiation effects in children born of exposed Marshallese parents cites Neel: "There is no evidence that genetic change was induced in children born of the exposed Marshallese children any more than there is unequivocal evidence of damage in the children born of the exposed Japanese."[185]

While exposure to ionizing radiation has long been suspected to increase mutation load in humans, scientific findings on atomic bomb victims from research by James V. Neel and others seems not to have shown significant genetic defects. However, new research emerging from post-Chernobyl studies shows that small

---

Robert Conard notes, "During the first four years the exposed women showed some increase in miscarriages and stillbirths. About 41 percent of the births in that period ended in nonviable babies compared with only 16 percent in the unexposed group." With reference to genetic effects, Conard reports, "We have to expect that there are genetic mutations that exist in these people." R. A. Conard, *Proceedings in the Second Interdisciplinary Conference on Selected Effects of a General War, October 4–7, 1967* (Princeton, NJ: Defense Atomic Support Agency, 1969), vol. 2, session 3: 118, 120, http://worf.eh.doe.gov/data/ihp1c/0282_a.pdf (accessed April 1, 2008).

[184] See Hardin B. Jones, *A Summary and Evaluation of the Problem with Reference to Humans of Radioactive Fallout from Nuclear Detonations* (Berkeley: University of California Radiation Laboratory, January 14, 1957). The question of the genetic effects of radiation exposure has been a heavily contested one. James V. Neel, the researcher involved in the first human subject experiment in the Marshall Islands, was also the principal architect of the Atomic Bomb Casualty Commission studies, including a five-year study of pregnant women in Hiroshima, Nagasaki, and Kobe (the "control" population situated 18 miles from Hiroshima). That study found evidence of miscarriages and congenital birth defects in all populations examined but identified the consanguineal "breeding pattern" of the Japanese as a significant factor in explaining the rate of birth defects in this population. The application of consanguinity as a discount factor assumed that marriages between first, second, and third cousins would produce higher rates of defect. Rates of defect were defined in relation to each of the three study groups. At the time, data to contextualize the Hiroshima/Nagasaki experience with the national Japanese experience, or with previous preradiation experiences, did not exist. In addition to problems in interpreting findings, a 1956 National Academy of Sciences review of the study identified serious problems with the definition and use of Kobe as a control population, recognizing significant problems with exposure determinations and acknowledging that the control population had also received exposure to radiation. See, J. V. Neel, "A Study of Major Congenital Defects in Japanese Infants, *American Journal of Human Genetics* 10 (1958): 398–445; and *Critique of the Report of the National Academy of Sciences: The Biological Effects of Radiation* (Washington, DC: National Academy of Sciences, 1956), 4, http://worf.eh.doe.gov/data/ihp1c/0507_a.pdf (accessed April 1, 2008).

[185] Dr. James Livermore, letter to Congressman Balos, Majuro, Marshall Islands, October 1976, http://worf.eh.doe.gov/data/ihp1a/2180_.pdf (accessed December 17, 2007). Neel bases this statement on work with Japanese survivors of the Hiroshima and Nagasaki bombings and

doses of radiation do produce genetic effects on human populations. A study published in May 2001 showed that children born to cleanup teams from the1986 Chernobyl nuclear plant accident in the former Soviet Union had an unexpectedly high genetic mutation rate when compared with siblings born before the accident. Children born to cleanup team families now living in either the Ukraine or Israel, and conceived after parental exposure to radiation, were screened for the appearance of new fragments using multisite DNA fingerprinting. Siblings conceived before exposure served as internal controls. External controls (nonexposed families) were also included in the study. The study recorded a sevenfold increase in the number of new bands in children born after their parents' Chernobyl exposure compared with the level seen in controls. A strong tendency for the number of new bands to decrease with elapsed time between exposure and offspring conception was established for the Ukrainian families. These results indicate that low doses of radiation can induce multiple changes in human germ line DNA.[186]

Regardless of the scientific arguments over statistically reliable indications of radiation-induced intergenerational effects, the Rongelapese have certainly experienced significant problems with pregnancy and abnormal births:

Almira Matayoshi: *There are so many kids who weren't properly formed when they were born. Many of the kids were tiny—very short. Kids like Mike, Julie, Carol, Kimo, and Harry. Some of these kids just lie there. They don't know how to move, like Kimo and Carol. There was another kid who would bite trees, crawl, and touch fires. He didn't know the difference. All the kids who were retarded died. These kids weren't alive for the bomb, but they had so many problems. We want DOE to take care of our kids, but they won't, even though they have so many problems from the radiation.*[187]

Dorothy Amos: *Some of the children who were born were deformed. Their arms and legs are short. Nothing covering their brains. Still others were like grapes. But some were not*

---

their offspring conducted for the Atomic Bomb Casualty Commission. In the 1960s, Neel and others conducted a restudy of the pregnancy outcome data, reporting findings of no significant genetic effects except changes in the sex ratio and blood protein. The restudy involved the same use of a consanguinity discount ratio and employed the same assumptions of no radiation exposure in the control population. Over the years, Neel's findings were cited when the Japanese and the Marshallese expressed concern over the intergenerational effects of radiation exposure. Neel later served as chair of the committee charged with assessing the status of Rongelap, reported in the 1994 National Research Council report *Radiological Assessments for Resettlement of Rongelap in the Republic of the Marshall Islands*. It is important to note that scientific findings emerging from Chernobyl exposures conclude significant genetic effect from exposure to low-level radiation. See H. Sh. Weinberg et al., "Very High Mutation Rate in Offspring of Chernobyl Accident Liquidators," *Royal Society Proceedings: Biological Sciences* 268, no. 1471: 1001–05.

[186] The study was conducted by researchers from the Institute of Haifa in Israel and the Academy of Medical Sciences of Ukraine and was published in the May 22, 2001, issue of the *Royal Society Proceedings*. See Weinberg et al., "Very High Mutation Rate."

[187] Almira Matayoshi, interview by Holly M. Barker, Honolulu, June 13, 2001.

the [offspring of] poisoned people, but they were the ones who went back in 1957, and after some years they also gave birth to those kinds. Doctor Conard said: "Sometimes there will be different things in your stomachs because you were exposed." There were lots of words like these words. . . . Some kids go right into the fire or slam into you—they're missing something. One kid they tied him up but he died because he crawled into the ocean and drowned. He just died on Mejatto. He was older, around ten years old. He looked okay, but something was wrong with his brain. His parents aren't the bomb people, but they lived on Rongelap after 1957. . . . Some kids live for a week or so. You see their brains. Their faces are okay, but their brains are scary. Their hands and feet come out of their torsos. They know how to breastfeed. They appear to be people, but they are different. We believed what DOE said, that some of our kids would be born differently. Many children were this way. . . . We came together to bury them because we knew and expected these kinds of kids. They gave birth on Ejit. It was like what, like grapes. What kind of children were those that appeared? One kid born on Rongelap, his feet were flipped up so the soles faced up. His head was big and soft. A different kind of child. My eldest daughter might have a heart problem. Another child has a bone problem—the bone sticks out of his back. One extra finger was also removed.[188]

One physician found a connection to reproductive problems and past exposure: "Adult Rongelap women, 16 to 34 years old in 1988, are more likely to have reproductive problems such as spontaneous abortions, still births or infant deaths if their parents were in the DOE 1954 exposed or comparison groups (post 1957 exposed)," compared to young women whose parents were not.[189]

Medical studies found "about twice as many abnormally terminated pregnancies" in the Rongelapese between 1957 and 1963.[190] Medical survey scientists told the Rongelapese that unsuccessful pregnancies should be expected.[191] Abnormal outcomes were documented, although no systematic study was conducted. The community reported significant reproductive problems:[192]

---

[188] Dorothy Amos, interview by Holly M. Barker, Ebeye, March 18, 1999.

[189] Rosalie Bertell, Committee on Interior and Insular Affairs, 10.

[190] Reported in Henry Kohn, *Rongelap Reassessment Project Report* (Berkeley, CA: Rongelap Reassessment Project, 1989), 14, http://worf.eh.doe.gov/data/ihp1d/400005e.pdf (accessed April 1, 2008).

[191] Timako Kolnij and Dorothy Amos, interviewed by Holly M. Barker, Ebeye, March 18, 1999.

[192] According to Helena Alik, secretary to Trust Territory district administrator Oscar DeBrum, all the medical files, records, and photographs of the deformed children and medical problems of the Marshallese were kept in a locked safe in DeBrum's office. One night, some unknown person opened the safe and burned all the contents. Medical files also disappeared from hospital facilities. Alik was also one of the control patients when the Rongelapese were on Ejit and she was in Majuro. At one time, her medical record at the hospital was huge, because she was involved in so many tests and procedures. Now her chart is almost empty. She doesn't know who removed the contents of her file. Interview by Holly M. Barker, Majuro, March 13, 1998.

Catherine Jibas: *I returned to Rongelap . . . in 1957. . . . It was around this time that I had my first pregnancy. My baby had a very high fever when he was delivered, and the attending health assistant conveyed his doubts as to whether my son would survive the night. He was so dehydrated from the fever that his skin actually peeled as I clasped him to me to nurse. The only thing we knew to do was to wrap him in towels. And so it was that I held him to my body throughout the night, changing the towels and willing him to fight for his life. He lost the fight just as dawn broke.*

. . . My second son, born in 1960, was delivered live but missing the whole back of his skull—as if it had been sliced off. So the back part of the brain and the spinal cord were fully exposed. After a week, the spinal cord became detached and he, too, developed a high fever and died the following day. Aside from the brain deformity, my son was also missing both testicles and a penis. He urinated through a stumplike thing measuring less than an inch. The doctors who examined him told me that he would not survive. And sure enough, he was dead within a week. You know, it was heart wrenching having to nurse my son, all the while taking care his brain didn't fall into my lap. In spite of his handicaps, he was healthy in every respect. It was good he died, because I do not think he would have wanted to live a life as something less than a human.

. . . The health assistant who delivered the child sent a message to Kwajalein, and I am certain those [U.S.] doctors came for the express purpose of seeing firsthand a live nuclear baby [ajiri ebaam]. In fact, they flew in the very same day the message was sent. . . . They did a complete physical [of the baby], took blood samples, and lots and lots of photographs.[193]

It is clear that at least some U.S. government researchers were interested in the abnormal births of Rongelapese women like Catherine. After the Marshallese health officer learned of Catherine's abnormal child and informed the U.S. doctors, the U.S. doctors dispatched him from Ebeye to see the baby:

Ezra Riklon: *I wrote to Dr. Conard about this. Because they asked me that—if there was any—they even tell me to talk to the magistrate there if there were any abnormal babies. They should notify me personally so that I can go there and do whatever I can to study these babies. And I report only one abnormal baby. After she was born, she was born—the baby was very funny looking. The legs and arms were there, but they were kind of larger than normal, and shorter than normal. You can see the body, but there was no skull—and there was no skull except a membrane of the brain, but you can see the brain with your own eyes, you can see the brain is moving. And the baby, the heart was beating also . . . and the baby was quite shorter than normal. Kind of thick and big, you know. . . . But you cannot see the skull. . . . This baby, I wrote a complete letter to Dr. Conard about what I found during*

---

[193] Catherine Jibas, interview by Holly M. Barker, Majuro, August 23, 1994.

my physical examination. I took a picture of the baby in different positions and sent it to Dr. Conard in Brookhaven.[194]

Ezra Riklon's 1994 comments are substantiated by his August 17, 1960, letter to Dr. Conard:

On August 16, 1960, 24 yrs. old Catherine gave a normal birth to a baby monster. Miss Catherine, as you probably know, is listed among the unexposed group. . . . The physical examination of the new born baby revealed a baby monster possessing a trunk but only an imperfectly developed head, from which a large part of the brain and skull is lacking. The testicles and the scrotum were absent. The penis appeared smaller than normal. Large upper part of the brain and meninges were visible due to lack of cranial vault. The child was still alive and breathing normally during my examination, but unable to cry when pinched hard with a sharp needle. The best and closest diagnosis that I could think of is Hemicaphalus or Hemicephalic Monster. . . . Lots of pictures of different positions of the Monster were taken and blood specimens. I sincerely hope that you will get them soon.[195]

Linguistic evidence indicates that the reproductive problems experienced by the Rongelapese women did not occur before the testing program and their exposure to radiation. Instead of using Marshallese words to describe the illnesses, the interviewees apply descriptive terms from their local environment when talking about these birth anomalies. They use words such as *octopus, grapes, hermit crab,* and *dog* to describe the deformed children. If these reproductive problems had existed before the testing program, they would have had proper Marshallese names, as do other illnesses experienced by the Marshallese before the testing program, such as normally occurring *jibun,* or stillbirths.[196]

Almira Matayoshi: *One person gave birth to something that looked like a dog.*[197]

Nerja Joseph: *In the years afterwards, one kid was born with a head like an octopus. Other kids never really grew and were very short.*[198]

---

[194] Ezra Riklon, interview by Holly M. Barker, Majuro, August 18, 1994.
[195] Ezra Riklon, letter to R. Conard, August 17, 1960. On file at the Embassy of the Republic of the Marshall Islands, Washington, DC.
[196] Holly M. Barker, "Fighting Back: Justice, the Marshall Islands, and Neglected Radiation Communities," in *Life and Death Matters: Human Rights and the Environment at the End of the Millennium,* ed. Barbara Rose Johnston (London: AltaMira Press, 1997).
[197] Almira Matayoshi, interview by Holly M. Barker, Honolulu, June 13, 2001.
[198] Nerja Joseph, interview by Holly M. Barker, Majuro, March 8, 1999.

Aruko Bobo: *I know a boy, actually a young man now, whose head is so large that his body is unable to support it, and his only means of getting around is to crawl backwards, dragging his head along. Like the movements of a coconut or hermit crab.*[199]

Reproduction problems and birth defects are now so common among the Rongelapese that every interviewee was able to describe problems suffered by their offspring. Every woman interviewed complained about experiencing reproductive problems. The Rongelapese believe these problems are directly linked to both the parents' and the children's exposure to radiation:

John Anjain: *There were sixteen kids DOE found with low blood-cell counts. Some kids died, some kids have severe problems. Like Julie, whose mouth kept growing bigger. There were nineteen kids that moved off of Rongelap. Only two or three are alive now. Those who are alive have problems.*[200]

The accounts by the Rongelapese correspond with Dr. Rosalie Bertell's testimony to the U.S. Congress regarding observations from her medical examination of the Rongelapese. In 1988 the percentage of Rongelapese adults with medical problems was 88.5 percent for men and 88.6 percent for women, and "among the children who were evacuated from the Rongelap atoll in 1985, there is a very high degree of ill health, with about 42% having medical problems. Medical problems which were identified only among the evacuated children were: multiple organ systems malfunctioning, autism, anemia, arthritis, arthralgia, epilepsy, Down's Syndrome, facial asymmetry, loss of nasal bridge, and meningitis. Heart disease was diagnosed in 9.2% of the evacuated children."[201]

Brookhaven scientists also documented a range of immediate and longer-term health consequences from radiation exposure, although they typically did not pass along their findings to the Rongelapese. For example, their research led to the conclusion that fetal or infant life was expected to be most sensitive to radiation exposure, but they did not notify the Rongelapese of their concerns.[202]

Assumptions that radiation-related health problems were limited to those resulting from acute exposure and immediate effects of a specific event meant that Brookhaven scientists failed to recognize some of the more systemic medical

---

[199] Aruko Bobo, interview by Holly M. Barker, Ebeye, August 27, 1994.

[200] John Anjain, interview by Holly M. Barker, Ebeye, March 16, 1999.

[201] Bertell, Committee on Interior and Insular Affairs, 10.

[202] See, for example, the Brookhaven medical survey coordinated by Robert Conard and the long-term study work conducted as part of this effort by W. A. Sutow. While data on the long-term effects of radiation exposure was published at times, research findings were not made available to the affected population until some twenty years into the study, following local protests and international scrutiny.

consequences of radiation exposure. For example, Brookhaven scientists assumed that the thyroid was resistant to radiation. This assumption translated to a failure to look for and treat thyroid abnormalities in the Rongelapese.[203] Over time, researchers observed "markedly stunted growth" in Rongelapese children and suggested possible thyroid deficiencies.[204] Follow-up research found "unquestionable damage to the thyroid gland, especially to those exposed below the age of 10" and found that "nearly all the people exposed on Rongelap to fallout at less than ten years of age have developed nodules (benign tumors) of the thyroid."[205]

Almira Matayoshi: *We didn't have any retarded or short kids before the bomb.*[206]

Isao Eknilang: *When I heard about the deformed kids, particularly the ones in my family, I felt afraid and furious. I just wanted to cut them (the U.S. government representatives) up with a knife.*[207]

Timako Kolnij: *One of my kids is mentally challenged. He was born in 1963. The child didn't walk until he was five years old. . . . He stays with me. He gets mad sometimes. I will have to care for him because he has to stay with me. He gets upset and agitated when I do something he doesn't like. He went to school for the challenged. He still hasn't got to the Department of Energy doctors because they say they don't take care of our children.*[208]

Problems associated with reproductive health, congenital deformities, and degenerative health conditions have been all the more difficult to endure given difficulties in accessing health care. After the community was returned to Rongelap in 1957, a local health assistant attended to their medical needs, with the exception of one yearly visit by Brookhaven National Laboratory. When Brookhaven doctors did arrive, they looked for specific signs of radiation illnesses they expected to see and ignored the complaints of the people (such as illness from consuming contaminated foods). During the fifty-one weeks of the year that Brookhaven was absent, the health assistant had to deal with unique and complex medical care needs. As noted earlier, a 1957 document from Jorulej Jitiam reveals his concern about problems he did not know how to treat, his fear that radiation was the cause of these problems, and a request for a trained doctor to visit Rongelap immediately.

---

[203] ACHRE, *Final Report*, 589.
[204] Henry Kohn, *Rongelap Reassessment Project Report* (Berkeley, CA: Rongelap Reassessment Project, 1989), 14, http://worf.eh.doe.gov/data/ihp1d/400005e.pdf (accessed April 1, 2008).
[205] Ibid., 14–15. Special Joint Committee Concerning Rongelap and Utirik Atolls, *A Report on the People of Rongelap and Utirik Relative to Medical Aspects of the March 1, 1954 Incident: Injury, Examination and Treatment* (Saipan, Mariana Islands: Congress of Micronesia, February 1973), 35, http://worf. eh.doe.gov/data/ihp2a/0328_a.pdf (accessed April 1, 2008).
[206] Almira Matayoshi, interview by Holly M. Barker, Honolulu, June 14, 2001.
[207] Isao Eknilang, interview by Holly M. Barker, Honolulu, June 13, 2001.
[208] Timako Kolnij, interview by Holly M. Barker, Ebeye, March 18, 1999.

According to an independent physician who visited the community in 1980, the Rongelapese were justified in their health care complaints. A July 15, 1980, report from Dr. Reuben Merliss (accredited by the AEC to use radioactive isotopes in medical practice for treatment of thyroid disorders) to Gordon Stemple, an attorney for the affected communities, gives a sense of the degenerative effects produced by acute and long-term exposure to fallout in Rongelap. Merliss notes the condition of Rongelapese subjects:

- Vision: "remarkable frequency of visual difficulties"
- Thyroid: "The frequency of thyroid tumors was shown by the examination of a number of patients who demonstrated thyroidectomy scars." There is an "epidemic" of thyroid disease. People ran out of thyroid replacement drugs. Exams show many thyroid cancers, in conflict with what BNL doctor Conard states. "There appears to be little doubt that the tumors, benign or malignant, are radiation-induced. There are just too many of them to be anything else. Otherwise one would have to postulate that the Marshallese had a remarkably high incidence racially of tumors of the thyroid, this existing before 1946, and the old people I spoke to denied this. They denied that prior to the bombs there was any particular epidemic of lumps in the neck."
- Reproductive problems: There are an "unusual number of stillbirths or the birth of monstrosities."
- It is clear that there is a spread of radiation from island to island by birds and fish.
- Diabetes: "When I spoke to the old people who remember the way the islands were before the nuclear testing, they all routinely deny that diabetes was a great problem for the inhabitants."
- Impotence: "There are some sexual problems among the males of the island, or among the females. A number of men from one atoll had told me that they developed a failure of sexual interest after the explosions, this persisting, and in several cases their families did not expand after the bomb blasts." "The testicles are in an exposed position, particularly in people who so commonly sit on the ground or squat as do the people of the Marshall Islands."
- Hypertension: High incidence—5 percent in average white American males but possibly up to 40 percent in the Marshall Islands. There is a link between "inadequate space or, inadequate food, competition between them for sustenance and living space," and hypertension.

Conclusion

The follow-up care of patients with total ablation of the thyroid also appears inadequate. Some had stopped their medicine because of side reactions and appear hypo-thyroid in my eyes. No regular follow-up has been pursued to allow change of thyroid medication, or increase or decrease of dosage. A general feeling of distrust of the Treaty Trust and Atomic Energy Commission physicians is wide-spread among the Marshallese. The people of the Marshall Islands I spoke to have no great faith in these physicians, do not consider them devoted to their interests, but instead representing the interests of the Atomic Energy Commission or the Trust Treaty authorities, and were reluctant to place their health in their hands. . . . I am also impressed by the failure of the physicians to communicate findings and prognosis to the people of these islands. Each patient is entitled to have his questions answered. He should be told the nature of the lesion discovered, and, if he asks for it, a prognosis should be given. The doctor should, when he can, inform the patient of the cause of his illness. These basic rights of a patient have been in large part ignored in the Marshall Islands, and I found very few Marshallese who were acquainted with the nature of their pathology.

Finally, Merliss noted need for "an improvement in diagnosis of hitherto unrecognized food-chain radiation caused diseases"[209]

In addition to feeling angry about their treatment as subjects rather than patients, the Rongelapese experienced profound humiliation, marginalization, and stigmatization as a result of their injuries and reproductive problems. These experiences occurred when people sought medical care with DOE doctors in public hospitals, were summoned for their exams by Brookhaven, sang in public, or even walked down the street:

Norio Kebenli: *Our radiation exposure was so embarrassing. Whenever we went to the hospital to get any kind of care, people would always point at us, they knew who we were. Many people used to say things like: "Don't marry the Rongelapese because they are sick and your kids will be sick." When people heard that the Utrikese were sent back to their islands from Kwajalein, people said that the Utrikese were sent back because they were getting radiation from the Rongelapese. The embarrassment still continues today. Last week during graduation, many kids from Rongelap graduated [from high school], but when they were asked where they are from, they didn't say Rongelap.*[210]

Kobang Anjain: *As a kid it was particularly bad when the AEC used to come and pull us out of school. I went to high school at ECES, a Christian school. The AEC would come and pull me out of my classroom for my exams. The kids made fun of me, and they would say: "Etal bwe jenaaj radioactive ippem" [Get out of here because we're going to get radiation from you]. I could never blend in because the kids always remembered what*

[209] Reuben Merliss, report to Gordon Stemple, July 15, 1980, http://worf.eh.doe.gov/data/ihp1a/3081_.pdf (accessed October 25, 2007).
[210] Almira Matayoshi, interview by Holly M. Barker, Honolulu, June 13, 2001.

happened to me, especially when the AEC turned up at our school. I lost friends because of what happened.[211]

Almira Matayoshi: *People didn't want to shake our hands for fear we would contaminate them. We were embarrassed to walk around where there were other Marshallese, because they would say things like "Rej kamour kiraap" [They give birth to grapes].*[212]

SUMMARY OF CONSEQUENTIAL DAMAGES

Living in an isolated, contaminated setting meant dealing with the pain, suffering, and hardships of a contaminated and poisoned food supply; fear and anxiety of additional exposures; sociocultural and economic damages associated with loss of access to the material means to sustain a healthy way of life; and the psychosocial trauma of living life with radiogenic illnesses. The people of Rongelap were further burdened with difficulties of caring for the degenerative conditions associated with radiation exposure; the pain and suffering associated with miscarriages and the birth of congenitally deformed children; the difficulties of raising physically disabled children and caring for increasingly feeble elderly; the fear and anxiety of additional exposures; the fear and anxiety of the intergenerational and other unknown effects of radiation exposure; and the psychosocial humiliation, marginalization, and stigmatization experienced by the population as a whole as a result of their nuclear victimization. The U.S. government ignored complaints about the medical monitoring program and routinely downplayed the effects of radiation through the use of a faulty control group.

Human Subject Research Experiences

Policies and priorities established in 1954 by U.S. scientists and military officials emphasized monitoring and human subject research rather than treatment of the Rongelap people. These policies and priorities shaped the nature of medical attention in subsequent years on Rongelap. In 1954 the Rongelapese were told that medical survey "treatments" were important and necessary to their health. However, in 1954 and in the years since, numerous procedures and exams were conducted, and samples obtained, for a wide range of experiments that had little or nothing to do with individual treatment needs.[213]

---

[211] Kobang Anjain, interview by Holly M. Barker, Majuro, June 13, 2001.

[212] Norio Kebenli, interview by Holly M. Barker, Honolulu, June 13, 2001.

[213] For example, see descriptions of the Thyroid Reserve Study and the Iodine Protein Study of different ethnic groups involving the administration of I-129. Robert A. Conard, *Protocol for*

In addition to documenting the health effects of human exposure to nuclear weapons testing, U.S. government scientists purposely exposed the Marshallese to additional radionuclides. ACHRE confirmed the existence of at least two radioisotope studies in which Marshallese citizens were used as human subjects in research without their awareness or consent. Documents released since the 1995 publication of ACHRE final report provide further evidence of human subject abuse.

In one study reviewed by ACHRE, a chelating agent (ethylene diamine tetra-acetic acid, also known as EDTA) was administered to Rongelap subjects seven weeks after exposure to radiation from weapons tests. EDTA was normally administered shortly after internal radiation contamination to remove radioactive material. Given the delay in administering the chelating agent, there was virtually no therapeutic benefit envisioned. The stated purpose of this experiment was to mobilize and detect radioactive isotopes already present in subjects' bodies as a result of their exposure to Bravo.[214] There were also adverse health risks associated with this experiment; in the early 1950s, several deaths occurred from kidney toxicity after EDTA treatment (the dosage used at that time was about ten grams per infusion. The current recommended dosage is three grams).[215]

In a second experiment discussed by ACHRE, chromium-51 (Cr-51), a radioactive tracer, was given to ten people resettled on Rongelap in 1957. The purpose of this experiment was to determine whether anemia observed among the Marshallese was an ethnic characteristic or due to radiation exposures. Researchers argued that since these people had not been present during the Bravo test, they were a viable control group that was "not exposed" to radioactive contamination. Chromium-51 was given without explanation or informed consent to tag red blood cells and measure

---

the 1974 Medical Survey in the Marshall Islands (Upton, NY: Brookhaven National Laboratory, 1974), document 0410764, 3–4, http://worf.eh.doe.gov/data/ihp1c/0764_a.pdf (accessed April 10, 2007).

[214] ACHRE, Final Report.

[215] Health concerns associated with chelation therapy are listed at Holistic-online.com. EDTA therapy poses a threat to patients who are elderly, have low parathyroid activity, or are suffering from heavy-metal toxicity. Heavy metals damage the kidneys, and too rapid infusion can overload them. Heavy metals likely to produce kidney damage include lead, aluminum, cadmium, mercury, nickel, copper, and arsenic. Other health effects include a rapid drop of calcium in blood levels, causing cramps and convulsions, and a drop in blood glucose, leading to insulin shock. "Chelation Therapy," Holistic online.com, http://www.holistic-online.com/chelation/chel_safety.htm (accessed April 1, 2008).

red-blood-cell mass.[216] ACHRE noted that in this experiment, "the tracer dose used would have posed a very minimal risk, but it was clearly not for the benefit of the ten subjects themselves."[217] This conclusion of minimal risk was based on ACHRE's assumption that the subjects were nonexposed Rongelapese.

In actuality, the Cr-51 received by the ten "nonexposed" Rongelapese subjects represented an additional burden to the already-considerable exposure from consuming contaminated foods and living in a radioactive environment. Furthermore, the 1959 medical survey plan suggests that Cr-51 was administered to a "control" group and to an "exposed" group, totaling twenty Rongelap subjects in this experiment. Review of declassified documents raises the possibility that the Cr-51 experiment was also repeated on at least two other occasions, in 1964 and 1965. Laboratory reports filed with Dr. Conard report Marshall Islands blood-volume data findings from twenty-one Rongelapese samples taken in 1964 and eighteen Rongelapese samples taken in 1965.[218]

The 1959 medical survey research plans also included instructions for the survey team to conduct additional human experiments and procedures:

• Study thyroid uptake by giving Iodine 131 to 25 to 30 people being tested for serum iodine values. Give small amounts of iodine 131 and 18 hours later use whole body counter to determine thyroid uptake.
• Conduct whole body gamma spectroscopy on as many of the group being examined as possible. Plant, soil, marine, and urine samples may also be analyzed in conjunction with University of Washington group.

---

[216] At the July 1954 AEC Division of Biology and Medicine conference "Long Terms Surveys and Studies of Marshall Islands," participants discussed and agreed on the need to use the Rongelapese in radioisotope studies that were unrelated to their exposure needs. Participants made specific reference to Cr-51 studies (Cohn, *Conference on Long Term Surveys,* 175–76):

Dr. Bugher: Have you thought much of the possibility of tagging procedures to measure red blood cell formation rates?

Cdr. Cronkite: We thought about it.

Dr. Bond: We thought about it, and thought it could be done, but wondered what the value of the program would be and whether we were justified in doing it.

Cdr. Cronkite: The problem of a questionable nutritional status and various things that can interfere with iron uptake that would be unconnected with exposure to radiation and the difficulty to get a truly unbiased random sample of normals for comparison. I don't know whether it ought to be done or not . . .

Dr. Dunham: I think if it is not controlled, it will not be meaningful. If it is controlled, it is more significant than bone marrow biopsies as far as the red cell cycle is concerned.

[217] Advisory Committee on Human Radiation Experiments, *Human Radiation Experiments,* 592.
[218] L. M. Meyer, handwritten note to W. E. Siri. Subject: Blood Plasma Data of Marshallese Subjects, http://worf.eh.doe.gov/data/ihp2/4625_.pdf (accessed October 25, 2007); L. M. Meyer, memo to W. E. Siri. Subject: Blood Volume Data of Marshallese Subjects (Enclosure: Assay of Tritiated Plasma), http://worf.eh.doe.gov/data/ihp2/4609_.pdf (accessed October 25, 2007).

• Give one 5-grain tablet of ferrous sulfate daily to 34 people. Comparison will determine whether iron deficiency is a factor in lowered hematocrits. Serum iron used on those not receiving iron tablets.[219]

In addition to experiments conducted during medical survey trips, Rongelapese underwent experiments during visits to U.S. research labs and hospitals. An April 9, 1965, letter from Dr. Leo Meyer to Dr. W. Siri reports findings from the administration of tritiated water (a form of the radioisotope tritium) and Cr-51 to six people who had traveled from Rongelap to Brookhaven labs.[220] Tritium was used to measure total body water, a factor of interest in determining the relative ability of the stomach to secrete iodine, bromine, and other halogens.[221]

In 1999 the DOE acknowledged that from 1961 to 1963, BNL conducted clinical tests involving Cr-51 and tritiated water on twenty-eight Marshallese subjects, including six subjects involved in both Cr-51 and tritiated water studies. This acknowledgement was in response to a specific request by the Republic of the Marshall Islands for information on Cr-51 and tritiated water studies conducted in the early 1960s. The DOE did not disclose information about Cr-51 or tritiated water tests that might have occurred in other years, nor did it disclose information about participants in other radioisotope studies, such as the I-131 thyroid-uptake studies, Sr-85 studies, and Fe-55 studies mentioned in medical survey planning documents.

The Rongelapese also underwent genetic studies. In 1956 the chairman of the AEC, Lewis L. Strauss, reported: "With reference to the Rongelap people, I have been advised that genetic studies will be undertaken at the time of the next and succeeding resurveys of the medical status of these people."[222] Rongelapese were also the subjects of chromosome abnormality studies. Samples collected during the 1964 survey found: "A total of 51 persons has been examined and although findings are not spectacular, it appears that there is a difference between the heavily exposed population and the controls."[223]

---

[219] R. A. Conard, letter to C. L. Dunham. Subject: Survey Which Involved the Use of Small Trace Amounts of Radioactive Materials (Upton, NY: Brookhaven National Laboratory, January 16, 1959), http://worf.eh.doe.gov/data/ihp1b/7783_.pdf (accessed April 1, 2008).

[220] Dr. Leo Meyer, letter to Dr. W. Siri, April 9, 1965, http://worf.eh.doe.gov/data/ihp2/4614_.pdf (accessed December 17, 2007).

[221] James S. Robertson, oral history interview by Michael Yuffee and Prita Pillai, U.S. Department of Energy, January 20, 1995, http://hss.energy.gov/healthsafety/ohre/roadmap/histories/0478/0478toc.html (accessed December 17, 2007).

[222] Lewis L. Strauss, chairman of the AEC, letter to Dr. Failla, October 12, 1956. On file at the Republic of the Marshall Islands Embassy, Washington, DC.

[223] Hermann Lisco, letter to George Darling, director of the Atomic Bomb Casualty Commission, April 29, 1966. ABBC box 22, file Marshallese Study (AEC) with ABCC Correlation, 1955–1971, National Academy of Sciences Archives.

Exams also involved procedures that were difficult to endure and at times undertaken for broader research purposes rather than monitoring and thus treating the health and well-being of the subjects. In at least one instance, Rongelapese subjects endured whole-body irradiation to help scientists calibrate instruments between Brookhaven and Argonne labs, as illustrated in the January 20, 1995, oral history testimony of Dr. James S. Robertson:

Q: *When was the first group [of Marshall Islanders] brought to Brookhaven for the first studies there? Or, if you can't remember, what was the purpose for bringing them to Brookhaven?*

Robertson: *Well, for some of those body measurements we wanted to [establish absolute values by calibrating them with other instruments]. Actually, they stopped at Argonne [National Laboratory, outside Chicago] and they were measured there before they were at Brookhaven. There were more sophisticated ways to measure things in the laboratory than what we could take with us to the island. I don't know all the motives for doing it. I think the main thing was to get better quality data.*

Q: *Were any studies done on the people [who] came from the Marshall Islands? Any tracer studies or anything of that nature?*

Robertson: *I don't think so. Some of them that were brought back were developing thyroid problems, and this was studied, but I wasn't directly involved in that. Again, the main thing that we were interested in was calibrating different instruments that were involved in the field and the laboratories, using [the islanders] as standard subjects and cross calibrating between what Argonne would determine as the body content and what Brookhaven would determine as the body content.*[224]

Beginning in 1954, medical survey scientists collected teeth from the people of Rongelap and Utrik. Trust Territory health staff were also instructed to collect deciduous teeth of children and to extract teeth in support of strontium-90 studies. Brookhaven National Laboratory has charts showing whether the teeth belonged to exposed Rongelapese, children of exposed Rongelapese, matching controls, or children of matching controls. The teeth were put in bottles and shipped back to New York for assessment of radiological content. In some cases, good teeth—not just decayed teeth—were removed.[225]

---

[224] James S. Robertson, oral history interview by Michael Yuffee and Prita Pillai, U.S. Department of Energy, January 20, 1995, http://hss.energy.gov/healthsafety/ohre/roadmap/histories/0478/0478toc. html (accessed December 17, 2007).
[225] Untitled document in the DOE declassified boxes showing which teeth were pulled from which exposed and unexposed people. On file at the Embassy of the Republic of the Marshall Islands, Washington, DC.

The assault received from radiation injuries was all the more difficult when compounded by the lack of information provided to subjects and treatment efforts that produced unanticipated harm. A 1978 letter from Dr. Conard to Dr. Walter Wyzen at the DOE discussed the case of "a complete thyroidectomy at the U.S. Naval Hospital in Guam. During the surgery, the parathyroid glands were inadvertently removed. The parathyroidectomy presents a more serious condition than the thyroidectomy, requiring a more precise and strict treatment regimen with lack of treatment resulting in more serious consequences."[226] According to Rongelapese informants, two girls, Chiyoko Tamayose and Lindy Nitihara, had their parathyroids mistakenly removed.

Lindy Nitihara described the trauma of being alone during her thyroid surgery, as well as medical problems resulting from the surgery:

Lindy Nitihara: *They sent me and Tija to the naval hospital in Guam. I was eleven years old, and I didn't speak English. There was no one with me to translate or explain what was going on. The rubber band on my underwear snapped, and there was no one there to even help me mend my one pair of underwear. The people laughed at our clothes and made fun of us there. I was terrified without my mother or father, or anyone. I think the surgery I had was bad because of the problems I've had, but I don't know, because up until now there is no medical record of what they did to me! I've asked DOE for my medical record, and they said there isn't one.[227]*

The medical survey experience was a dehumanizing experience in which the Marshallese were transformed from individual patients with specific medical needs into subjects whose identities were reduced to patient numbers. In 1963 the AEC made identification cards for each person in the study.[228] As in 1954, when Project 4.1 began, people were asked to remove their clothes for identification photos. People did not understand why they needed identification cards or why they had to remove their clothes:

Lijon Eknilang: *I don't know about the men, only about the women, but when it came time to get our identification cards and numbers from the Atomic Energy Commission, we had to hold up our individual numbers on a card and have our photographs taken. What I don't understand, and still don't understand to this day, is why we had to take our clothes off just to have our ID pictures taken. One by one, each one of the women was told to disrobe and have her photograph taken.[229]*

---

[226] Dr. Conard, letter to Dr. Walter Wyzen, 1978, http://worf.eh.doe.gov/data/ihp2/1464_.pdf (accessed October 25, 2007).
[227] Interview by Holly M. Barker, Honolulu, June 13, 2001.
[228] John Anjain, record book, 1957, 97.
[229] Interview by Holly M. Barker, Majuro, March 28, 2001.

From the onset of the medical survey, Marshallese patients complained that doctors were much more interested in studying them than in addressing their medical concerns. Informants noted that they would complain to the doctors about problems, but the doctors ignored their complaints and focused on their thyroids, blood, or urine. The doctors dehumanized the people, treating them as mere identification numbers or body parts and showing no concern for their feelings or sense of humiliation. Informants were especially traumatized by repeated instances of being photographed naked:

Almira Matayoshi: *There were at least three times when they took pictures of us naked — at the school on Rongelap and at the Majuro hospital. They had us take off every single piece of clothing and stand naked before their cameras. We were afraid. They told us to take off our clothes. I protested and didn't want to take off my underwear, but they said I had to. We had no choice. Men and women lined up for pictures, as we were told. Everyone knew what was happening. This humiliation was so powerful to me. We thought we were getting doctoring, but look what they did instead!*[230]

Further evidence of the dehumanization of the patients is evident in the descriptions of the care the people of Rongelap received. Routinely, the doctors violated Marshallese custom (by having people undress or discuss illnesses in front of relatives), humiliated people (by pulling students out of classrooms for medical exams), failed to explain or translate procedures and prognoses to patients, failed to provide patients with their medical charts, and sent them to unknown doctors and facilities in the United States and Guam for unexplained procedures:

Lijon Eknilang: *In 1981 three women from Utrik and I went to Brookhaven. It was so cold at the time because there was lots of snow outside. During this visit, we were given mammograms for the first time. We didn't understand what the doctors were doing to us or why, and it was embarrassing to have them touching our breasts and putting them in the machines. There was no translator for us.*[231]

Ellen Boaz: *I had a tumor. How many months and weeks they shocked my head in Washington [D.C.]! They examined us . . . we don't know what it is they did to us or what they saw.*[232]

Efforts by the people of Rongelap to exercise their rights to meaningful informed consent were dismissed, dissuaded, or met with trivial efforts to placate concerns.

[230] Almira Matayoshi, interview by Holly M. Barker, Honolulu, June 13, 2001.
[231] Lijon Eknilang, interview by Holly M. Barker, Majuro, March 28, 2001.
[232] Ellen Boaz, interview by Holly M. Barker, Mejatto, August 26, 1994.

Medical survey reports describe efforts to entice Rongelapese participation in exams and procedures with offers of hats, money, movies, ice cream, and cigarettes. Sometimes researchers used more abusive means. Helena Alik, a participant in the control group, notes that Trust Territory police sometimes physically removed her from school or home to ensure her participation in medical survey exams and procedures that she did not understand and that the AEC doctors did not explain.[233] John Anjain records in his magistrate's book that when he and others from Rongelap and Utrik traveled to the United States in 1957 to be examined, AEC doctors told them that if they spoke with anyone, they would be left behind forever in Hawaii.[234] In this case, U.S. government researchers used intimidation to ensure that no one questioned their treatment of Marshallese patients.

SUMMARY OF CONSEQUENTIAL DAMAGES

Exams, procedures, and sampling typically occurred without meaningful informed consent. Some procedures resulted in the mistaken removal of parathyroid glands. Often procedures were painful, resulted in additional exposure to radiation, were dehumanizing, and had little or no connection to the health treatment needs of the individual. While the U.S. government acknowledged culpability and accepted limited responsibility for specific radiation-related health effects, it systematically ignored the broader constellation of health needs that were directly and indirectly related to exposure. Decades of medical focus on research priorities rather than individual health care needs constituted systematic and long-term abuse of the Rongelapese—abuse that continues to generate hardships for individuals, families, the broader communities of Rongelapese, and the nation.

Evacuation of Rongelap in 1985

Despite assurances by the U.S. government that Rongelap was safe, the Rongelapese believed that their health and safety were further compromised after 1957 by living in an environment contaminated by radiation. As detailed above, the food and water on Rongelap gave people blisters in their mouths and made them sick. Interviewees believe that their environmental exposure led to premature deaths and severe deformities in their offspring.

In 1982 the U.S. government published two reports that would ultimately lead the people of Rongelap into self-exile in 1985. The first report was a Defense

---

[233] Interview by Holly Barker, Majuro, March 13, 1998.
[234] John Anjain, record book, 1957, 97.

Nuclear Agency document demonstrating that the U.S. government had been aware of the wind shift toward Rongelap before it detonated the Bravo shot. The second report was a DOE survey of the northern Marshall Islands. In this report, the DOE concluded that Rongelap was safe for habitability. Unfortunately, this report "failed to include the information on health and medical history of the residents of Rongelap. Issues of risk to children, to pregnant women and to those in ill health were not addressed."[235] The report did not refer to data on individual body burdens from the whole-body counting done by the Marshall Islands Radiological Safety Program; these data were classified and were released only as part of a Clinton administration study on the use of human subjects in radiation research.

Former senator Jeton Anjain (brother of John Anjain) of Rongelap disagreed with the DOE's notion of habitability. He countered the DOE definition of habitability with one "defined in Black's Law Dictionary as: Condition of premises which permits inhabitants to live free of serious defects to health and safety."[236] Believing lives to be in danger from environmental sources of radiation on Rongelap, Senator Anjain worked with the RMI Nitijela, or parliament, to pass a unanimous resolution asking the U.S. government to relocate the Rongelapese. The U.S. government ignored this request. In frustration, the Rongelapese turned to Greenpeace, an environmental nongovernmental organization, for assistance. In May 1985 Greenpeace dispatched its ship *Rainbow Warrior* to Rongelap Island to assist the Rongelapese with their evacuation. The Rongelapese moved to Mejatto Island, a small island on the western side of Kwajalein Atoll. The U.S. Department of Energy declared that there was "no justification" for the self-evacuation of the Rongelapese.[237] The Rongelapese viewed Mejatto as a temporary relocation while waiting for independent scientific work to be conducted. The move was extremely difficult; people left with fear for their health and safety and feared that, once gone, they would never be able to return.[238]

Rokko Langinbelik: *We left Rongelap because we didn't want our children to be poisoned like we are. Even if we were sad, we left. We left because we care about our children.*[239]

Isao Eknilang: *The throats of people from Rongelap exploded [from grief], and I, my throat, was decimated the most because of my land.*[240]

---

[235] Anjain, Committee on Interior and Insular Affairs, 10; Bertell, Committee on Interior and Insular Affairs, summarized in Mindy Palleija, *Summary of Hearing Before the Subcommittee on Insular and International Affairs* (Department of Energy, November 16, 1989), http://worf.eh.doe.gov/ihp/chron/G70.pdf (accessed April 1, 2008).

[236] Anjain, Committee on Interior and Insular Affairs, 16.

[237] Ibid., 14, 16.

[238] Henk Haazen and Bunny McDiarmid, *Report on the Marshall Islands: 1/4/86–1/8/86* (Auckland: Greenpeace, 1989).

[239] Rokko Langinbelik, CMI Nuclear Institute presentation, Majuro, May 7, 1999.

[240] Isao Eknilang, CMI Nuclear Institute presentation, Majuro, May 7, 1999.

The Marshallese say that their emotions are in their throats rather than in their hearts. Literally, Isao Eknilang considers the loss of his land to be an explosion of his heart.

SUMMARY OF CONSEQUENTIAL DAMAGES

The Rongelapese decided to leave their homes because of fear and concern for their safety. Environmental contamination from the nuclear weapons testing program forced them to abandon all that was familiar to and owned by them. Some of the consequences of this involuntary displacement include the hardships of struggling to survive on other people's land; loss of the means to reproduce sociocultural traditions; and erosion of trust regarding U.S. government efforts to disseminate information and deliver medical care.

Current Conditions Endured by a Fragmented Rongelap Community

After leaving Rongelap in 1985, the Rongelap community dispersed to Mejatto, Majuro, Ebeye, Hawaii, and, more recently, the mainland United States. Involuntary resettlement placed hundreds of people on small bits of rented land, creating extremely dense, unsanitary, and impoverished communities.

In urban areas, on the limited land in Mejatto and in the Rongelap community in Hawaii, people can no longer get food by growing, harvesting, and utilizing the wealth of their local environment as they did on Rongelap. Now they have to purchase the vast majority of their food. The Rongelapese complain that it is exceedingly difficult to feed their families. The custom of feeding and helping others is beginning to disappear, because people no longer have the means to care for others. Furthermore, when the Rongelapese were moved off their homeland, they did not have the education or job skills necessary to seek employment and earn the money needed for food purchases.

Approximately 350 people moved from Rongelap to Mejatto in 1985. Mejatto is a small island approximately 1 mile long. It is situated on the western end of Kwajalein Atoll, approximately 60 miles from Ebeye Island, the nation's second largest urban area. Mejatto has no airstrip. Its residents are completely reliant on small boats for travel to Ebeye.

Because Mejatto was not previously inhabited by a full-time residential population, when the Rongelapese arrived they had to plant their primary food crops: pandanus, breadfruit, and coconut. These plants took five years to mature and bear fruit. Dangerous tides and rocky reefs around Mejatto made it difficult for Rongelapese men to fish and to pass on knowledge about fishing to the younger generation. The islands surrounding Mejatto had food, but the Rongelapese were reluctant to gather it because they had no rights or permission to use these islands.[241] Field trip boats serviced the island with supplies once every three months. The people did not have motorboats or money for gas to make the ten-hour trip to Ebeye Island over rough seas.

The interviewees clearly describe many of the difficulties in securing food, transportation, and housing on Mejatto:

Nerja Joseph: *On Mejatto, there was no food when we got there. We had to plant coconuts, pandanus, and other foods.*[242]

Timako Kolnij: *People literally starved on Mejatto. There was no transport for food, no government support for food.*[243]

Ken Kedi: *When we left Rongelap, we were told everyone would have big houses on Mejatto. We only got small houses, and the guys had to sleep outside under the trees, where they were rained on.*[244]

These problems made people long for their home atolls, which were too contaminated to occupy:

Johnsay Riklon: *Mejatto is really different from Rongelap. Mejatto is really small. There are scarce resources, such as fish and grown foods. In Mejatto, people fish only two to three times a week, but on Rongelap people fished every day. People depend on USDA [U.S. Department of Agriculture] and imported foods. It's like a camp. People are not healthy. They have bad diets, diseases, and the health services are inadequate. It's expensive for me to visit Mejatto. . . . It is too rough to fish in Mejatto, and there are too many sharks. People die in the current around Mejatto.*[245]

Boney Boaz: *I didn't want to leave Rongelap in 1985, but the elders did, so I went. Because Rongelap was no good, I left, but I didn't like Mejatto. It is pretty difficult to fish on Mejatto. Wind prevents people from fishing. On Rongelap, there were many choices: many ways to fish and many places to fish. . . .*
    Kids on Mejatto don't know how to sail an outrigger canoe. It's too low to launch there. On Mejatto it's hard to sail and hard to teach.

---

[241] Henk Haazen and Bunny McDiarmid, *Report on the Marshall Islands* (Auckland: Greenpeace 1986), 3. On file at the Nuclear Claims Tribunal, Majuro, RMI.

[242] Nerja Joseph, interview by Holly M. Barker, Majuro, March 8, 1999.

[243] Timako Kolnij, interview by Holly M. Barker, Ebeye, March 18, 1999.

[244] Ken Kedi, interview by Holly M. Barker, Barbara Rose Johnston, and Stuart Kirsch, Majuro, March 1, 1999.

[245] Johnsay Riklon, interview by Holly M. Barker, Barbara Rose Johnston, and Stuart Kirsch, Majuro, February 28, 1999.

On Mejatto there is lots of American food. There's more food, local food on Rongelap. There was nothing at all on Mejatto in the beginning. Now it's better because we planted food.

Mejatto is bad, because it's hard to get back and forth. It's dangerous on all the small islands. Some people have disappeared, some drowned. It's hard to move about.[246]

According to the Greenpeace team that helped evacuate the Rongelapese to Mejatto, teenage boys had a particularly difficult time with the transition to Mejatto: "They have little to do. . . . They sleep, talk story, play guitar and softball and now and then go fishing or help with community jobs. . . . They are in between cultures, exposed to the American way of life through TV and videos and the consumer goodies available in Majuro and Ebeye. . . . Outer island life is very different and does not prepare people for the materially orientated imported lifestyle."[247]

The Rongelap community on Mejatto suffers from problems of isolation and lack of meaningful infrastructure. In March 1999, while conducting research on sociocultural values of land, the research team tried unsuccessfully for many days to rent a boat to travel to Mejatto. The fact that we were unable to secure a boat despite money and connections illustrates the transportation difficulties that the Rongelapese on Mejatto continue to encounter. The Rongelapese complain that it is difficult to get supplies and people back and forth between Mejatto and Ebeye, and it is expensive, often costing as much as one thousand dollars a trip.

The Rongelap community on Ebeye suffers from the immense problems of urbanization. Ebeye is the most densely crowded area in the Pacific Islands, with a population of approximately fifteen thousand people on one-tenth of a square mile of land. Housing, water, electricity, employment, education, health care, shade, and play areas are all difficult to come by.[248]

Interviewees on Ebeye complain of their inability to take care of their needs without money, overcrowded housing, ill health, unemployment, lack of knowledge about Rongelap among youth, and the burial of Rongelapese on other people's land:

Jerkan Jenwor: *What is life like now? It's filled with sickness. It was better on the island [Rongelap]. I came to Ebeye because my kids have to go to school. Food is hard when you don't live on your own land. It is also hard to find enough money to pay for school tuition.*[249]

Timako Kolnij: *Now we stay at one of the typhoon houses at Dump Town. Six of our kids are married. There are lots of grandchildren. There are thirty some people in three rooms.*

---

[246] Boney Boaz, interview by Holly M. Barker, Ebeye, March 17, 1999.

[247] Haazen and McDiarmid, *Report on the Marshall Islands,* 3.

[248] For a characterization of living conditions on Ebeye, see Howard French, "Dark Side of Security Quest: Squalor on an Atoll," *New York Times,* June 11, 2001, 3.

[249] Jerkan Jenwor, interview by Holly M. Barker, Ebeye, March 17, 1999.

You would laugh if you could see us sleeping—everyone together. I really need a house. My husband is retired, but he still works, because no one else has a job.[250]

Isao Eknilang: *I have a house on Majuro, but it's not the same. I don't know the people. My freedom is gone. Gone also is the practice of taking food to people. Now it's "kwe wot kwe im na wot na" [I take care of myself and you take care of yourself].*[251]

Rokko Langinbelik: *Marshallese custom used to make kids behave, but now they don't respect us. Now that we're gone from Rongelap, all of our most precious customs are gone. When we were on Rongelap, we practiced them all.*[252]

Malal Anjain: *My father died. There are five people in my family now. No one in my family works. We survive just from our quarterly payments [from the Rongelap trust fund]. It was better when my grandfather was alive. My stepfather was good, too. He died at Christmas. I want to go to college, but it will be tough on the people at my house, because no one works.*[253]

When the Rongelap community left their homelands, they left a subsistence lifestyle. They entered a monetary economy without the skills or training to find and secure employment. Yet the Rongelapese cannot live on Ebeye or Majuro without money. They must purchase everything they need:

Mwenadrik Kebenli: *On Ebeye . . . we buy things. We don't have pandanus leaves to make sleeping mats. I lie on the tile. We need money for everything. When it runs out, there is no more food. Our things to cook with break. I still haven't eaten breakfast today [at 11:15 a.m.]. There's no medicine. There's no vehicle to go and get birds to eat. They are far away. . . . My children, they grew up on Ebeye. They just hang around.*[254]

Burying family members is extremely difficult because of limited land, and because people believe their spirits will not rest properly unless they are on their own land:

Timako Kolnij: *Even if you move to a [new] place, you still remember your true place. It's not good to die away from our land. We shouldn't have to ask permission to be buried [on other people's land]. Even if we are far away, we know the place where we should be together and where we share lands that we should bury our loved ones together on.*[255]

[250] Timako Kolnij, interview by Holly M. Barker, Ebeye, March 18, 1999.
[251] Isao Eknilang, interview by Holly M. Barker, Honolulu, June 13, 2001.
[252] Rokko Langinbelik, interview by Holly M. Barker, Honolulu, June 14, 2001.
[253] Malal Anjain, interview by Holly M. Barker, Gugeegue, March 19, 1999.
[254] Mwenadrik Kebenli, interview by Holly M. Barker, Rongelap, March 16, 1999.
[255] Timako Kolnij, interview by Holly M. Barker, Ebeye, March 18, 1999.

Lijon Eknilang: *We shouldn't have to ask to be buried. Here on Ebeye, it's too crowded to bury. We have to look for places to bury our dead. I want to go back to Rongelap to die. Now they're starting to cremate on Ebeye because there's no place to bury now. They told us it's too full to bury. The cemetery is full. Burial areas are overcrowded, but cremation is against the custom and religion.*[256]

Obtaining access to adequate health care in all the Rongelap communities has been a major problem since initial exposure. After the evacuation from Rongelap in 1985, medical surveys continued to visit the dispersed community twice a year, which meant that for fifty weeks of the year, people did not have access to adequate medical care for their unique and complex needs. When the Compact of Free Association came into effect in 1986, the 177 Health Care Program (177 HCP) provided for the routine needs of some patients. But the program is open to all members of the Enewetak, Bikini, Utrik, and Rongelap communities and is grossly underfunded (with only seven to fourteen dollars per patient per month to support all health treatment needs).

In 1998 the Department of Energy dropped its arrangement with Brookhaven National Laboratory. Pacific Health Research Institute (PHRI) assumed responsibility for DOE's health care program. PHRI is in the process of trying to give the Rongelapese full-time access to medical doctors, but this is forty-seven years after their initial exposure. Furthermore, the DOE mandates that PHRI treat only the radiogenic conditions of its patients, a decision based on DOE policy and not congressionally mandated law. Exceptions to this rule are made from time to time, when a patient's life is in danger from a nonradioactive illness, but are rare. Congress intended for DOE to provide for the radiogenic and nonradiogenic needs of the community, but DOE maintains that it needs clearly articulated congressional language to change its policies. Because DOE is unwilling to treat nonradiogenic illnesses unless they are life threatening, and because the 177 HCP lacks funding, patients complain continually about being ping-ponged back and forth between medical providers that are reluctant or unwilling to care for them.

Lost in this conflict over health care responsibility have been the individual needs of Rongelapese patients. Informants allege that a number of recent deaths occurred needlessly because Rongelapese people suffering from the various effects of their exposure were unable to receive adequate health care. As of this writing, September 2001, these deaths include George Anjain, Jabeo Jabeo, Joe Jitiam, Edmil Edmond, and Bweradrik Eknilang.

The Rongelapese also believe that their children have radiation-related health care needs. They point to the high number of miscarriages, stillbirths, deformities,

---

[256] Lijon Eknilang, interview by Holly M. Barker, Ebeye, March 18, 1999.

retardation, and stunted growth in the children of people who were exposed to Bravo or who lived on Rongelap after the testing program. DOE's mandate does not include the offspring of the Rongelapese, and, once again, the 177 HCP lacks the resources to adequately care for them.

Because health care for the Rongelapese is limited in the RMI, many Rongelapese have opted to move to Hawaii to try to get access to U.S. public health care facilities. In Hawaii people are burdened by the extremely high cost of living. The costs of renting apartments, purchasing food and clothing, and providing for the basic needs of families are exorbitant. As a result, people often live in crowded apartments to help make ends meet.

In June 2001 interviews with Rongelapese living in Hawaii, much of the conversation focused on the difficulties of negotiating medical bureaucracies and confronting the biases and stigmatization associated with their medical conditions. A number of informants noted that political officials in Hawaii characterize all citizens of the Freely Associated States as nuisances and drains on public resources. Officials do not understand the special medical needs of the Rongelapese who suffer the degenerative effects of exposure to near-lethal amounts of radiation. Conflicts over health care responsibility—who should pay for what services—leave the Rongelap community in Hawaii struggling to obtain medical care. Providers often refuse to care for Rongelapese patients for lack of payment on previous bills that they cannot and should not have to pay. Informants expressed their belief that all community members, including offspring, should have access to any and all medical care they need as a result of U.S. government actions. They are confused by the bureaucracy of the medical establishment in the United States and frightened about the consequences of their inability to pay their medical bills. Yet they find this confusion and fear preferable to living in the Marshall Islands, where there is extremely limited medical care available to them and almost none for their children.

Naiki Ribuka: *If you live in the U.S., or if you leave the RMI to get medical care, you're not eligible for any [radiation-related] assistance. All of the medical bureaucracy is hard to understand. We get billed for care, but we don't have any insurance or any means to pay for our medical care. But we need the care; what can we do?!*[257]

Rokko Langinbelik: *Our kids need medical care. If we die, who cares for them, because we know their medical needs are beyond those of other kids?!*[258]

---

[257] Naiki Ribuka, interivew by Holly M. Barker, Honolulu, June 14, 2001.
[258] Rokko Langinbelik, interivew by Holly M. Barker, Honolulu, June 14, 2001.

Informant testimonies from all segments of the Rongelap community suggest a fragmented group living in extreme poverty, struggling with the comprehensive assaults resulting from their exposure and loss of land. The disbursement of the community into several main residence locations means that the Rongelapese changed from a close community, in which everyone worked together, into a geographically fractured community. The physical disbursement makes communication and group cohesion difficult for the Rongelapese. Each of the different communities where the Rongelapese now reside presents challenges for the Rongelapese that they did not encounter on their own land.[259]

The consequential damages of this fragmentation are reflected in national reports and statistics. Extremely dense residential patterns created by communities of exiles living on rented land have created or exacerbated terrestrial and marine pollution. The impoverished condition of the Rongelap community has intensified local resource use (especially fishing, a common resource). The ecosystemic viability of "host" island environs has been dramatically degraded. The Rongelap community represents some 8 percent of the total Marshall Islands population. While an estimated 49 percent of the nation is able to support household needs through agricultural production (living off the land), 100 percent of the Rongelap community is alienated from land and the traditional resources needed for survival.[260]

The people of Rongelap also suffer sociocultural hardships. While it is clear that the Rongelapese community would have experienced changes in diet, lifestyle, and employment over time, the weapons testing program accelerated the pace at which people experienced these changes. People could not adapt to this accelerated rate of change. This loss of the means to be self-sufficient impacts the people of Rongelap in complex ways:

Kobang Anjain: *Now I have to buy food all the time. I have to listen to the landowners on the land where I stay. Even if I don't like what the landowners tell me to do, I have no*

---

[259] Johnsay Riklon, interview by Holly Barker, Barbara Rose Johnston, and Stuart Kirsch, Outrigger Hotel, Majuro, February 28, 1999.

[260] Data from figure 4.2: "Households Engaged in Subsistence Activities, 1988," *Statistical Abstracts of the Marshall Islands* (Majuro: Republic of the Marshall Islands). Some 48.77 percent of RMI households grew food; 69.04 percent fished for food; 45.76 percent kept livestock; 35.97 percent were involved in handicraft production for household use. The Rongelap community could gain access to fishing with boats, or with coastal access permission. However, the intense household density results in severe pollution of coastal and near-shore waters, habitat degradation, and localized overfishing.

choice, because it's not my land, and I can't do what I want to. It's not the fault of the alabs of the land where I stay now—we lost our place.[261]

Loss of land affects diet, health, and household economy and severely inhibits the Rongelapese ability to produce or reproduce cultural knowledge about the local environment—knowledge that is essential to the survival and long-term well-being of the community. Traditional and current use-rights arrangements include an implicit agreement to care for the land. The scant terrestrial resources in the overcrowded urban areas render it impossible for the Rongelapese to live a subsistence lifestyle or to provide the maintenance of the land that traditional rental rights require. The Rongelapese living in Ebeye and Majuro do not have rights to freely use without permission coconut, pandanus, and other material resources situated on the land, or reef heads and other marine resources adjoining the land.

The inability of the Rongelapese to access plant resources also means the loss of a free, culturally appropriate system of health care, as well as the cultural respect reserved for people who are knowledgeable about Marshallese medicine. Outer island populations rely heavily on traditional Marshallese medicine to assist them with their ailments.

Instead of traditional work on the land and in the sea, the Rongelapese are largely idle in their relocation. Suicide, malnutrition, alcoholism, smoking, and lack of physical fitness were complaints among the interviewees. The psychosocial damages associated with the hardships, anxiety, and reality of life for the people of Rongelap are reflected in apathy and misery in the younger generations:

Ken Kedi: *Here are the examples of youth activities for the Rongelapese: gather, talk, smoke, make yeast [a homemade alcohol]. Increasing suicide. A number of suicides have been attempted on Mejatto. One family [the Balos family] had three or five suicides.*[262]

Since moving off their land, the Rongelapese diet has changed substantially. In the place of arrowroot, breadfruit, and fish, the people now subsist primarily on rice, flour, and canned meats when they are available. Fresh produce from local grocery stores is prohibitively expensive for families on a limited income. Furthermore, the younger generation has become accustomed to processed foods because of the lack of availability of traditional foods. A traditional means of maintaining family ties

---

[261] Kobang Anjain, interview by Holly M. Barker, Honolulu, June 13, 2001.
[262] Ken Kedi, interview by Holly M. Barker, Barbara Rose Johnston, and Stuart Kirsch, Majuro, March 1, 1999.

had been the sharing of food. As indicated earlier, in urban areas, on the limited land in Mejatto, and in Hawaii, people no longer have the ability to get their foods locally:

Naiki Ribuka: *When we were on Rongelap, we would worry about everything that we ate, even if they said it was safe. We thought the foods were contaminated. Now we always struggle with money. We don't have enough money to purchase the food we need, we can't buy what we want. "Jej jojolaer. Ainwot bao ko rej jan bujik bwe ejjelok jineer." [We're lost and wandering. (We're) like the birds that cry out because their mother is gone.]*[263]

When living on other people's land, the Rongelapese can use only the rainwater they collect from their roofs in catchments. In some cases, people can get permission to use groundwater, but they have to ask permission to use any and all water they do not catch themselves. In recent years, the RMI has suffered severe droughts. People do not have the means to gather sufficient water for their families:

Abacca Anjain-Maddison: *As a child, I remember fighting to get the hose so I could get water on Ebeye. We would also fight with each other to get on the boat to go over to Kwajalein to get water.*[264]

Loss of land not only affects the means to sustain life but also severely impacts the traditions that accompany the end of life. The displaced Rongelapese have no connection to their ancestors. During a site visit to Rongelap in 1999 with the elders of the community, an elderly Rongelapese woman proceeded immediately to the cemetery after the airplane landed. She greeted the gravestones of her family out loud with "Yokwe kom," or "Hello to you." It is painful for the Rongelapese to be physically removed from their ancestors. Furthermore, it is not customary to bury the deceased on land that does not belong to the family:

Almira Matayoshi: *My father really wanted to be buried on Rongelap when he died, but there was no way to get him there.*[265]

Ken Kedi: *Most elders died. They died wanting to go back. My great-grandfather Talekerab said he wouldn't leave the island when it was time to evacuate. After they took him off, he started fasting and demanded to go back. Every day he would chant, sing, and talk of the*

[263] Naiki Ribuka, interview by Holly M. Barker, Honolulu, June 14, 2001.
[264] Abacca Anjain-Maddison, interview by Holly M. Barker, Honolulu, June 13, 2001.
[265] Almira Matayoshi, interview by Holly M. Barker, Honolulu, June 13, 2001.

good old days. He would ask: "Why are we here [in Mejatto]?" Because he said he wouldn't eat until he went back, he died.[266]

In addition, the Rongelap community has been deeply affected by social forces dividing people into groups of "exposed" and "nonexposed." They have been constantly challenged by the hardships of dealing with a range of degenerative problems resulting from their exposure to radiation. Thyroid disorders and cancer treatments affect people's voices, which makes evidence of exposure immediately apparent. The Rongelapese feel deep anguish at the changes in their singing voices as a result of their thyroid disorders. Singing is a vital part of Marshallese gatherings; all formal occasions, community gatherings, and church services are singing events. Singing and dancing are often related activities that transmit stories and legends, commemorate significant events, and communicate honor and respect. Numerous informants expressed deep sadness about their inability to sing in public and their embarrassment about the loss of range in their notes:

Norio Kebenli: *We used to love singing! Christmastime was always the most important to us because we come together and sing. Now no one is interested in participating in our traditional Christmas get-togethers anymore, because we can't sing. We've lost interest. People who used to sing well now have flat voices. Personally, I don't sing in public anymore because people stare at me. It's like we're in a constant state of puberty, where our voices keep cracking. I feel sorry for Jia Riklon. He can't speak unless he holds a contraption to his throat to make sound—he has only one level of sound he can make, so he certainly can't sing.*[267]

Almira Matayoshi: *After the bomb, we can't harmonize anymore. Everyone's voice is a base, and there are no more sopranos amongst us. We have no interest in singing anymore. People make fun of us when we do and say "Etiroit men ne" ["That thing near you is thyroid" or "That thing of a person has a thyroid problem"].*[268]

Ellen Boaz: *At the time when they cut my throat, I thought they—well, I don't know, I really can't sing anymore, but I want to sing again, but now I can't. I really don't know. After the people cut my throat, I really can't sing at all anymore. My voice won't go high anymore. Is that not from the contamination?*[269]

[266] Ken Kedi, interview by Holly M. Barker, Barbara Rose Johnston, and Stuart Kirsch, Majuro, March 1, 1999.
[267] Norio Kebenli, interview by Holly M. Barker, Honolulu, June 13, 2001.
[268] Almira Matayoshi, interview by Holly M. Barker, Honolulu, June 13, 2001.
[269] Ellen Boaz, interview by Holly M. Barker, Mejatto, August 26, 1994.

SUMMARY OF CONSEQUENTIAL DAMAGES

The loss of customary lands and the means to sustain community cohesion, coupled with degenerative disease associated with radiation exposure, has produced distinct effects on diet, health, and household economy. Extremely dense residential patterns created by communities of exiles living on rented land have created or exacerbated terrestrial and marine pollution. The impoverished condition of the Rongelap community has intensified local resource use, and the ecosystemic viability of host island environs has been degraded. Restricted access to critical resources inhibits the ability to teach younger generations about the means to sustain a self-sufficient way of life. Loss of access to customary lands further inhibits efforts to transmit key information across the generations—knowledge that is essential to the survival of the community if it is ever to return to customary lands. Some hardships are endured by individuals, such as the many Rongelapese who experienced physical pain and suffering as a result of having their thyroid glands removed or treated, continue to endure the discomforts and side effects of taking lifelong thyroid medication (nausea, lethargy or hyperactivity, hot and cold flashes), or mourn throughout their life the loss of a unique part of their personal and social identity: their ability to sing. Other hardships are endured by families, the community, and the nation, as they also struggle to care for children who were born retarded or handicapped and an elderly population with severe degenerative health care needs. These social, cultural, economic, and environmental problems are linked to exposure and the loss of lands and are consequential damages from nuclear testing. These hardships are exacerbated by the fact that virtually all of the Rongelapese community—including those alive during the testing program, those who lived in the highly contaminated environment, and their offspring—have significant health care needs related to their exposure and struggle with inadequate access to meaningful health care. Yet there is no medical program to adequately address these needs. Patients are frustrated and angry that the U.S. government has failed to provide for them and has ping-ponged them back and forth between inadequate programs.

Part 4

Summary of Damages, Needs, and Compensation Concerns

Claims by the People of Rongelap for Hardship and Related Consequential Damages of the Nuclear Weapons Testing Program

Evidence in parts 2 and 3 of this report support the following findings:

- The people of Rongelap experienced involuntary displacement from Rongelap and Ailinginae atolls when they were physically removed from their atolls (March–May 1946; 1954–1957). They lost access to a viable healthy ecosystem (and thus were displaced from their ability and right to safely live in their environment) when they were returned to their atolls (1957–1985). They became exiles (1985–present) when they were finally informed of the life-threatening contamination levels in their homeland.
- Families were deprived of their right to live and use lands on Rongerik Atoll. Rongerik Atoll was taken for U.S. naval use following World War II and used as a weather and fallout tracking station during the nuclear testing program (1946–1958). The U.S. Navy, without getting permission or providing compensation, used Rongerik as a resettlement site for the Bikinians (1946–1948). In 1957, when the Rongelap community was resettled on Rongelap and Ailinginae, all access to and subsistence use of Rongerik Atoll was prohibited by the United States due to severe contamination from nuclear weapons fallout.
- Exposure concerns involve much more than the exposure to radiation and fallout from a singular testing event in 1954. Exposure concerns involve the persistent presence of contamination from sixty-seven atmospheric tests of nuclear weapons in the Marshall Islands. This contamination includes radioactive elements

173

released through nuclear explosions, as well as tracer chemicals, such as arsenic, used to "fingerprint" the fallout from each weapon. The people of Rongelap, Rongerik, and Ailinginae were exposed to external radiation and other toxic substances not only from fallout but more significantly from internal ingestion: breathing dust and smoke from household and garden fires, drinking water, consuming terrestrial and marine food sources, and living in houses and using material culture fashioned from contaminated materials.

- Exposed people of Rongelap include those living on Rongelap and Ailinginae in 1954 who were exposed to Bravo and other test fallout; those who were re-settled in 1957; those who were born on the contaminated atoll; those who were exposed to materials and food originating from Rongelap, Rongerik, and Ailinginae atolls; and the descendents of people exposed to radioactive contaminants. Given the synergistic, cumulative, and genetic effects of long-term exposure to radioactive isotopes and other environmental contamination from military testing, exposure is of concern to this and future generations.

- The people of Rongelap, Rongerik, and Ailinginae, along with other Marshallese, served as unwitting subjects in a series of experiments designed to take advantage of the research opportunities accompanying exposure of a distinct human population to radiation. Human subject research involving the Marshallese was initially funded by the AEC in 1951 in an effort to document "spontaneous mutation rates" to better estimate the genetic effects of radiation produced through the nuclear weapons testing program. Research on the human effects of radiation was intensively conducted beginning in March 1954, with efforts to document the physiological symptoms of U.S. servicemen and Marshallese natives exposed to fallout from the Bravo test. Initial findings from this and other biological research projects helped shaped the goals of and approach to an integrated long-term study on the human and environmental effects of nuclear weapons fallout that began in 1954 and continued through 1997.

- The people of Rongelap believe, and the documentary record confirms, that the United States was aware of the extraordinary levels of fallout from Bravo and subsequent tests and continuing levels of radioactivity, was aware of contamination in the marine and terrestrial ecosystems, was aware of the bioaccumulative nature of contamination, noted radiation-induced changes in vegetative and marine life that islanders relied upon for food, monitored the increased radiation burdens of the people returned to Rongelap in 1957, and documented the human health consequences of this systematic and cumulative exposure. Medical exams, especially from the 1950s to the early 1970s, were surveys meant to document bioaccumulation processes and the physiological symptoms related to radiation exposure rather than clinical efforts to treat the various health conditions of the people of Rongelap. Periodic medical surveys also subjected the people of Rongelap to various procedures that produced biological samples in support of a wide range of experiments, many of which had little or no connection to the individual health and treatment needs of the people of Rongelap. Varied human subject experimentation also occurred during medical treatment trips to research laboratories based in the United States and Guam. Ethnographic and documentary evidence demonstrates that these human subject experiences were painful, abusive, and traumatic.

- In addition to biophysical injuries, exposure to the environmental hazards generated by the U.S. nuclear weapons testing program (and related biomedical research) resulted in stigmatization and other psychosocial injuries that adversely affected individuals, the community, and the nation. Nuclear testing introduced new taboos: certain lands and foods were off-limits; marriage to certain people involved new social stigmas; birthing presented new fears and health risks; family life often involved the psychological, social, and economic burden of caring for the chronically ill and disabled. The failure of the U.S. government to provide the Rongelap people with accurate information concerning environmental hazards and risks, coupled with contradictory pronouncements on what was and was not safe, created taboos that were incomprehensible yet dominated living conditions after the onset of testing in the Marshall Islands. This transformation in the loci of control over taboos from a Marshallese cultural realm to a U.S. scientific realm undermined the rules and customary power structures that shaped, interpreted, and reproduced strategies for living in the Marshall Islands. The fear of nuclear contamination and the personal health and intergenerational effects from exposure color all aspects of social, cultural, economic, and psychological well-being. This imposed stigmatization adversely affects the economy, society, family life, and individual health and well-being of the people of Rongelap, Ailinginae, and Rongerik, and to varying degrees the entire nation.

- After leaving Rongelap and Ailinginae atolls in 1985, the Rongelapese faced severe hardships as they struggled to rebuild some semblance of community in Mejatto, Majuro, Ebeye, and other locations. Involuntary resettlement placed hundreds of people on small bits of rented land, creating extremely dense, unsanitary, and impoverished communities. The Rongelap community represents some 8 percent of the total Marshall Islands population, and while an estimated 48 percent of the nation is able to support household needs through agricultural production, the Rongelap community is alienated from land and the traditional resources needed to survive. This loss of access affects diet, health, and household economy.

- Extremely dense residential patterns created by communities of exiles living on rented land has created or exacerbated terrestrial and marine pollution. The extreme impoverished condition of the Rongelap community has intensified local resource use, and the ecosystemic viability of host island environs has been degraded. Restricted access to critical resources inhibits the ability to teach younger generations the means to sustain a self-sufficient way of life. Loss of access to customary lands further inhibits efforts to transmit key information across the generations—knowledge that is essential to the survival of the community if it is ever to return to customary lands. These social, cultural, economic, and environmental problems of urbanization are linked to the loss of lands and involuntary displacement and are consequential damages from nuclear testing.

• Nuclear testing destroyed the means to sustain a self-sufficient way of life for the people of Rongelap. Customary uses of Rongelap, Rongerik, and Ailinginae atolls encompassed a rich range of social, cultural, and economic activities, values, and meanings that allowed a vibrant, marine-based, self-sufficient, and sustainable way of life. Current and customary laws, traditions, and subsistence production patterns involve an inherited system of rights to both terrestrial and marine resources. The consequential damage and injuries of contamination from the nuclear weapons testing program affects both terrestrial and marine ecosystems, including the natural and cultural resources that sustain life. Damage assessments and compensatory actions need to include consideration of lagoons, reef heads, clam beds, reef fisheries, and turtle and bird nesting grounds, as well as those resources important for sustaining the social and cultural aspects of life, including family cemeteries, burial sites of iroij, sacred sites and sanctuaries, and *morjinkōt* land. Because sustainable subsistence production requires access to multiple locations, so that people do not overuse the resource base, cleanup and resettlement of the main island of Rongelap Atoll is not sustainable without restoration of all Rongelap's islands and Rongerik and Ailinginae atolls. Compensatory actions should reflect a commitment to replace, restore, or create new means to sustain a self-sufficient way of life.

Nuclear testing created environmental hazards, health problems, hardships, and other consequential damages that will persist for decades to come. Compensatory actions should incorporate principles of nuclear stewardship and should provide sufficient funds, facilities, expertise, and training to give the people of the Marshall Islands the means and ability to conduct their own intergenerational epidemiological surveys, conduct environmental risk assessments, develop culturally appropriate environmental risk-management strategies (including monitoring contamination levels and decay rates and remediating terrestrial and marine ecosystems), and provide intergenerational medical care.

Consequences of These Events and Injuries

Stigmatization and Other Psychosocial Injuries

Exposure to the environmental hazards introduced by the U.S. nuclear testing program resulted in stigmatization and other psychosocial injuries that adversely affected the local economy, the community, and the nation. Customary rules and traditions are an essential component of the Marshallese psyche. Rules that restrict behavior (termed here as taboos) were traditionally defined and imposed by internal actors in ways that established and regulated social, political, and economic relationships. Traditional taboos made sense. Nuclear testing introduced new taboos: certain lands and foods were off–limits; marriage to certain people involved new social stigmas; birthing presented new fears and health risks; family life often involved the psychological, social, and economic burden of caring for the chronically ill and disabled. The knowledge of and control over what was and was not taboo resided in the hands of outsiders—U.S. government scientists and the agencies that sponsored and controlled their research. The failure of the U.S. government to provide the Rongelap people with adequate and accurate information concerning environmental hazards and risks, coupled with contradictory pronouncements on what was and was not safe, created taboos that were incomprehensible yet dominated all facets of society and way of life. This transformation in the loci of control over taboos from a Marshallese to a U.S./scientific realm undermined the rules and customary power structures that shaped, interpreted, and ensured the community's ability to care for itself. Furthermore, the fear of nuclear contamination and the personal health and intergenerational effects of exposure color all aspects of social, cultural, economic, and psychological well-being. This imposed stigmatization adversely affects the economy, society, family life, and the individual health and well-being of the people of Rongelap, Ailinginae, and Rongerik, and to varying degrees the nation as well. National economic development options and market response are intrinsically linked to the nation's nuclear history and the accompanying stigmatization of radioactive contamination.

The psychological impacts of the U.S. nuclear weapons testing program are severe. These impacts are not based on unjustified, unsubstantiated fears but result from decades of mismanagement of the radiation crisis by the U.S. government. Prior to the testing, Rongelapese survival depended on knowledge and understanding of the local environment. The Rongelapese knew when, where, and how to cultivate the resources necessary for survival. With the introduction of a hazardous, invisible threat to their environment and health, the Rongelapese feared the radiation they could not see or taste when they ingested it. In this sense, radiation has become the new taboo for the Rongelapese. The Rongelapese are forbidden to eat their traditional foods, such as the coconut crab, because of an intangible hazard. Their own lands have also become taboo, since they are off-limits because of radiation contamination. As a result, all members of the community experience a cultural agony from the dispossession and removal from land and culturally important places and the subsequent disintegration of community identity.[1]

---

[1] The psychosocial effects from invisible environmental contaminants have been well documented. See, for example, H. M. Vyner, *Invisible Trauma: The Psychological Effects of the Invisible Environmental Contaminants* (Lexington, MA: Lexington Books, 1988). See especially the section titled

The U.S. government's mismanagement of the Rongelapese radiation crisis exacerbates people's fears about radiation. The U.S. government prematurely declared an island safe and resettled the people there. As a result of the premature resettlement, the health of the Rongelapese was compromised. The U.S. government monitored increased radiation body burdens in the Rongelapese and failed to inform them about their risks of environmental exposure.[2]

The U.S. government also created conflicting directives for the resettled Rongelapese. For example, the people were told they could eat coconut crab, and then were told they could not. Finally, the Rongelapese believe the U.S. government is unwilling to take responsibility for their exposure because it will not discuss safety in terms of absolute safety. Instead of telling the Rongelapese definitively whether or not they are putting themselves at risk, U.S. government scientists discuss the issue in terms of calculated risks, whereby a few people could be affected, but each person has to decide whether or not to take that risk. The U.S. government's wrongful resettlement of the people, failure to disclose increased environmental exposure risk, and refusal to talk about safety in definitive terms increases the stresses and psychological burdens of radiation exposure for the Rongelapese people.

The Rongelapese have a firsthand understanding of the effects of radiation on human health and the environment. Their experiences differ from the scientific effects of radiation that the U.S. government imposes on the Rongelapese. For example, the scientific community failed to reach a decisive conclusion about the effects of radiation on the second generation. The Rongelapese have experienced changes in reproduction among women and changes in the health of their offspring since the testing program. These changes remain underreported, however, as cultural barriers prevent women from reporting or registering their experiences with either U.S. or Marshallese health authorities. The U.S. government's unwillingness to acknowledge the effects of radiation that the Rongelapese attribute to the testing program undermines the Rongelapese people's trust in the U.S. government's willingness to reconcile all the consequences of the testing program.

"Stigmatization and Other Psychosocial Injuries." In Australia, dispossession and removal from land contributed to a wide range of physical and psychological health effects, including spiritual issues of health. These have been documented in the National Aboriginal Health Strategy and are currently the focus of government compensation efforts. For an extensive discussion of the psychological consequences of exposure to the atomic bomb in Hiroshima, see Robert Jay Lifton, *Death in Life: Survivors of Hiroshima* (Chapel Hill: University of North Carolina Press, 1968). Lifton documents the psychohistorical impacts of radiation exposure, such as stigmatization, fear, health problems beyond those acknowledged by the U.S. and Japanese governments, and strain on family relations. Lifton also notes that these problems exist for Japanese born even after the two atomic bombs were dropped.
[2] See concerns raised by Tommy McCraw in his December 16, 1982, memorandum to James de Francis regarding a meeting on Majuro Atoll pertaining to the DOE northern Marshall Islands survey (http://worf.eh.doe.gov/ihp/chron/H7.pdf). It was at this meeting that the Marshallese were first informed that residence in Rongelap presented radiological risks to their health.

The ethnographic data collected for this project reveal the anxiety of the Rongelapese about the insidious, invisible radiation threat that has taken them from good health to illness, from their homelands to exile, and from trust to skepticism:

Nerja Joseph: *I won't go back. The radiation won't go away.*[3]

Dorothy Amos: *We were afraid of the powder and the explosions. We were afraid they would drop powder again.*[4]

Lijon Eknilang: *Psychologically, you stop believing in everything around you. Your feeling of safety no longer exists when the radiation-contaminated medicine and food around you is no good, restricted, and makes you worry.*[5]

Household Economic Injuries

As noted earlier, on Rongelap, Rongerik, and Ailinginae, the Rongelapese had sufficient resources to provide for the majority of their needs. Some provisions, such as kerosene, tin roofing, lamps, cigarettes, matches, sugar, rice, and flour, were purchased. The Rongelapese acquired these provisions by selling their natural resources to merchant ships that periodically visited their islands. Rongelapese men sold copra, a major source of income to the outer RMI atolls throughout the twentieth century. Women generated income from the sale of handicrafts or foodstuffs that they made individually or as part of a cooperative, the White Rose. Natural resources have value to the Rongelapese as items of exchange between families and neighbors; resources are exchanged for other resources, for labor, or to mark important cultural occasions. Today the Rongelapese live on small segments of other people's land, typically in dense clusters of shacks built from scavenged pieces of plywood, tin, and other materials. The houses have relatively little surrounding vegetation. The few resources that may be present (coconut, breadfruit, papaya, or pandanus trees) belong to the landowner and not people who live there. Environmental contamination has robbed the Rongelapese of their customary access to natural resources used to sustain households and communities, exchange for other goods, or generate an income.

Community Injuries

Leaving Rongelap Island has had profound implications on the Rongelapese community. For example, fragmentation has made it difficult for the Rongelapese

---

[3] Nerja Joseph, interview by Holly M. Barker, Majuro, March 8, 1999.
[4] Dorothy Amos, interview by Holly M. Barker, Ebeye, March 18, 1999.
[5] Lijon Eknilang, advisory committee meeting, Majuro, March 3, 1999.

to remain unified as a community. Because people reside in different locations, and because the younger generation grows up in a lifestyle radically different from that of their parents and grandparents, community unity is a greater challenge to the local leadership. As the former city manager for the Rongelap local government observed, "I am a city manager with no city," because there is no capital city for the Rongelapese people in their diaspora.[6]

While the fragmented community members still have many common interests, such as housing, health concerns, and the future of their land, the local leadership has to work harder to establish communication and a common plan of action to reach its objectives. In comparison, the Rongelapese community previously worked together constantly to ensure the survival and well-being of all community members on Rongelap, Rongerik, and Ailinginae atolls.

In addition to the problems caused by fragmentation, relocation from Rongelap has undermined cultural authority and respect for elders; people don't listen to their elders when they no longer fear the reproach of being kicked off their lands. Lacking land and resources, the people experience great difficulty generating tribute to offer to their iroij. Dispersed communities and high transportation costs mean that the iroij, *alabs*, and ri-jerbal have few opportunities to meet as a group. Birth, marriage, and death—the cycles of life in a community—are events no longer tied to the land and are celebrated without overt reaffirmation of land rights and respect for those who confirm land rights. Landholders and elders are experiencing an erosion of their ability to hold the community together, and an erosion of their own social power, since their authority is tied to use of the land and resources. The interviewees are clear about how displacement affects a sense of community:

Johnsay Riklon: *The community is scattered and splintered with no base. There are many different categories of Rongelapese now—those who have lived there, those who have not, teenagers, etc.*[7]

Wilfred Kendall: *Consider the role of individuals in a society defined by the environment. There are three tiers of rights. These rights are reciprocal. Alab and ri-jerbal roles are specific. Community life comes from the land. The movement of people from their land changes all of this: people are no longer connected to the land, or anything. When people are not on their land with its social ranks, they cannot perform what is expected of them. This makes them feel insecure. Land is a privilege, and it tells you who you are in the community. . . . The role of the alab and iroij are specific in nature. Moving away really messes up things.*

---

[6] Troy Barker, discussion with Barbara Rose Johnston at the Outrigger Hotel, Majuro, March 5, 1999.

[7] Johnsay Riklon, interview by Holly M. Barker, Barbara Rose Johnston, and Stuart Kirsch, Majuro, February 28, 1999.

Going to new land, new areas, you don't know what land belongs to whom. There is no iroij system there.[8]

John Anjain: *Sometimes today, there are problems determining who the alab is. There are more problems nowadays because there are more people, money is of greater importance, and because people marry with outsiders and mix the blood. When you marry Rongelap to Rongelap, it is in the best interest of the community to keep the conflicts down. Being moved off the land accelerates marriage to outsiders and reduces the power of landowners. It's easier for family to push you out, especially when your mother is from outside.*[9]

Displacement has resulted in the Rongelapese marrying people from other atolls. Traditionally, marriage within the Rongwelap community was preferred because it maintained ties to the land.

Political Impacts

The RMI national government has a constitutional responsibility to provide for the needs of the people and the protection of their environment. According to section 5 of the RMI constitution:

> No land right or other private property may be taken unless a law authorizes such taking; and any such taking must be by the Government of the Republic of the Marshall Islands, for public use and in accord with all safeguards provided by law. . . . [And] where any land rights are taken, just compensation shall include reasonably equivalent land rights for all interest holders or the means to obtain the subsistence and benefits that such land rights provide.

Despite these protections of rights by the RMI constitution, the RMI government is forced to reconcile the U.S. government's taking of private property from the Rongelapese. The RMI government lacks the human, financial, and institutional resources to provide the housing, food, water, and other needs that the Rongelapese used to provide for themselves on Rongelap, Rongerik, and Ailinginae. The RMI national government does not have three spare atolls to provide the equivalent of the land the Rongelapese lost. Nor does it have the resources to enable the Rongelapese to live self-sufficiently. Rongelapese dislocation causes a drain on national resources as the RMI government struggles to provide for the needs of the Rongelapese, especially their health care needs.[10]

---

[8] Willfred Kendall, advisory committee meeting, Majuro, March 2, 1999.
[9] John Anjain, interview by Holly M. Barker, Ebeye, March 16, 1999.
[10] Constitution of the Republic of the Marshall Islands, section 5.5.

The displacement of the Rongelapese from Rongelap, Rongerik, and Ailinginae also affects the political composition of the national government. Political representation in the nation's parliament, the Nitijela, is conditioned on physical occupation of the land. Each occupied atoll or large island, no matter how small its population, has the right to elect one senator. If there are no people residing on an atoll, there is no political representation for that area in the Nitijela. According to the former senator representing Rongelap Atoll, nuclear testing disrupted politics on the three atolls by reducing three atoll communities to just one:

Johnsay Riklon: *Ailinginae and Rongerik are large atolls, but they are treated like small islands by the United States. People should have been returned to these atolls, and there should be three senators representing the three areas, not one. In this regard, the testing affected the political makeup of the country . . .*

Now people talk about Rongelap as if it is just one place, Rongelap Island. But it was the fallout and testing that brought people from Ailinginae and Rongerik to Rongelap. Actually, there should be political representation from all three areas, since there are people with land rights in those different places.[11]

Rongerik Atoll lacks representation because the Rongelapese have been denied the right to return to their land. Following World War II, the U.S. Navy took control of the atoll, establishing a weather and fallout tracking station on Rongerik and, for a two-year period (1946–1948), using the atoll to host the displaced people of Bikini:

John Anjain: *We moved from Rongerik during the war to stay together on Rongelap. We didn't use it after the war, because the navy used Rongerik. People lived with their relatives on Rongelap, but the land rights were different than on Rongerik.*[12]

Ailinginae Atoll lacks representation because in 1957 the U.S. government confined resettlement to a single island in Rongelap Atoll and declared Ailinginae off-limits for permanent occupancy.

Stewardship Concerns

On the island of Rongelap, and on the small islands of Mejatto, Majuro, and Ebeye where the Rongelapese currently reside, the Rongelapese are unable to practice principles of responsible stewardship or easily transmit their knowledge

---

[11] Johnsay Riklon, interview by Holly M. Barker, Barbara Rose Johnston, and Stuart Kirsch, Majuro, February 28, 1999.

[12] John Anjain, interview by Holly M. Barker, Majuro, March 16, 1999.

of sustainable access and use to the younger generations. Notions of stewardship involve care of the land and limited use of resources in multiple marine and terrestrial locations. The inability of the Rongelapese to adhere to principles of stewardship while confined to Rongelap Island during their resettlement, and while currently living on Mejatto, Ebeye, and Majuro, means that the Rongelapese are contributing to the degradation of the land they use.

Because the Rongelapese live far from their traditional lands, no one is able to maintain the land and protect the resources and property for future generations. The Rongelapese express concern about people who sail to their now unpopulated atolls and pillage their resources:

George Anjain: *When land is evacuated, people can't protect their resources. Fishing boats and other boats steal the clams and other resources. The local government is unable to protect its resources. Foreign fishing and other vessels come and pillage the natural resources, such as the giant clams and turtles. Because there are no people living on the land, there is no one to protect the land.* [13]

Countless generations of future Rongelapese will inherit property contaminated by radiation. Unfortunately for the Rongelapese, the more time they are off the atolls, the more difficult it is for them to exercise their land rights and ensure that the land is passed on to future generations. Rongelapese who grow up away from Rongerik, Ailinginae, and Rongelap lack knowledge about essential cultivation areas and dangers that would be important to survival if they were to return to the land. Part of the difficulty of passing land on to future generations is that some of the books about the lands and wetos were burned or destroyed.

Mike Kabua: *People couldn't take all of their possessions with them [when they were evacuated]. This leads to disputes. People used to work by consensus.* [14]

Loss of Way of Life

Most informants emphasized that what has been lost is the means to sustain a way of life. Contamination forced the Rongelapese to abandon their three atolls. The Rongelapese must now rent or stay on other people's land. While the Rongelapese have places to live, they do not have just compensation for the U.S. government's taking or rental of their atolls. The U.S. government has not given them the means to rent other people's land. They do not have access to the natural resources they need as the "means to obtain subsistence and benefits" necessary for survival. [15]

---

[13] George Anjain, interview by Holly M. Barker, Ebeye, March 15, 1999.
[14] Mike Kabua, advisory committee meeting, Majuro, March 3, 1999.
[15] Constitution of the Republic of the Marshall Islands, section 5.5.

Compensation Concerns

"[A]ny Marshallese citizens who are removed as a result of test activities will be reestablished in their original habitat in such a way that no financial loss would be involved."

—"Petitions Concerning the Trust Territory of the Pacific Islands,"
United Nations Trusteeship Council, July 14, 1954

In the course of our interviews, one of the most common questions we heard from informants was: How can you possibly place a monetary value on land, or compensate for many lifetimes of suffering?

How can any amount of money ever compensate for the pain and loss experienced by those who lived through the testing, by those who endured years of living in a contaminated setting, by those unwitting test subjects in a series of radiation-exposure experiments, by those who struggle to retain identity and fulfill family needs while enduring the socioeconomic hardships of life on Ebeye, Mejatto, Majuro, and other places of exile? Land represents a perpetual resource— something that sustains this and future generations. How can a sum of money be equivalent?

For the Rongelapese, what has been lost is the means to sustain a self-sufficient way of life. Toward this end, we asked our informants to think beyond individual complaints and monetary awards and to imagine broader forms of compensatory action. What actions might help the Rongelap people and the nation reclaim a healthy, viable way of life? What kind of research is needed? What actions might help the displaced community regain its sense of integrity? Identity? Self-sufficiency? These questions prompted a wide range of ideas and concerns about how to proceed in the event that some compensation is granted. Many of these ideas are summarized below.

Land Compensation Concerns

Many responses involved compensation to current owners of weto rights. Informants explained that customary traditions and current practices involve a fluid rather than fixed system of rights to wetos. Weto rights are inherited, with a woman usually investing a male member of her family with the power to manage use rights. To claim the right to use a weto, one must know its boundaries and history. The power to recognize and validate that claim rests in the hands of customary authorities (*alabs* and iroij). Changes in family structures and changes in customary power structures (for example, the death of a mother, *alab*, or iroij) produce new distributions of use rights. Monetary compensation for denied use of or damage to land based on use rights for a fixed point in time imposes a system of individual property rights. Under customary practice:

[T]he Marshallese system of land tenure provides for all eventualities and takes care of the needs of all of the members of the Marshallese society. It is, in effect, its social security. Under normal conditions no one need go hungry for lack of land from which to draw food. There are no poor houses or old people's homes in the Marshall Islands. The system provides for all members of the Marshallese society, each of whom is born into land rights.[16]

Monetary compensation to a primary landholder for loss of use over a fixed amount of time undermines a very complex Marshallese land-tenure system. By imposing a simple compensation scheme that does not account for the multiple loss of use rights over time, this approach fails to recognize that "land in the Marshall Islands is placed in many categories, each with its own descriptive name and rules of inheritance. The Marshallese system of land tenure had developed to meet the needs of this particular group of people and is the dominant factor in the cultural configuration. Any radical change by outsiders would disturb the society and could do irreparable damage."[17] The fixed, Western notion of compensation threatens traditional, fluid systems of "reciprocal rights and obligations of all classes within the framework of the society"[18] and the rights of future generations.

In addition to failing to recognize the full range of rights use, fixed compensation approaches create vulnerabilities that vary according to gender, class, and age and exacerbate the displacement problems of the Rongelapese. Many informants observed that the increased incidence of lease payments to individuals (and other forms of payment to individuals, including compensation) does not result in the same patterns of distribution that customary payments in food products allowed. Monetary payments are typically transactions between individuals and are much less transparent than the public presentation of tributes and the redistribution of goods that occur in customary settings. Many informants expressed the need for compensatory and remedial actions that reflect community and intergenerational rights and needs.

Alienation of land rarely occurs in the Marshall Islands, because land is the lifeline for lineage and the basis of Marshallese culture and survival. As explained to project researchers in March 1999 by a member of the Land Value Advisory Committee:

Mike Kabua: *It is extremely rare for people to alienate their land. Traditionally, this only occurred when the iroij gave land as a gift or when it was won through war. Alienating land is so rare that there is no word for lease or borrowing land in Marshallese, because that was never done. Land was never leased to another Marshallese because land is so precious.*[19]

---

[16] Jack A. Tobin, *Land Tenure Patterns*, 1.
[17] Ibid., 2.
[18] Ibid., 4.
[19] Mike Kabua, advisory committee meeting, Majuro, March 2, 1999.

If compensation payment is awarded to individual owners of weto rights whose lands were irreparably damaged or lost, and money is exchanged for land, the land system will change in ways that cannot be predicted.

When people are forced to leave their land, the iroij lose their authority over the people who manage and work their land. The people lack access to the resources they are expected to provide to the iroij. Since the iroij distributes the resources among all people, the loss of resources from the Rongelapese reduces the iroij's ability to distribute goods to the *alab* and ri-jerbal on atolls other than Rongelap, Ailinginae, and Rongerik. This situation reduces the iroij's power and means that the people get less assistance:

> **Mike Kabua**: *Without people occupying the land, there is no iroij. The word* iroij *means "many people" and comes from the words* er woj. *What's the point of being an iroij without any land?*
>
> *When the bomb exploded, the culture was also gone, too. It is impossible for people to act in their proper roles. Our social roles are something you use every day, twenty-four hours a day: You have to use it every day or you lose it.*[20]

Traditionally, the care provided by the iroij was rewarded by work and gifts from the people under the iroij. Because the Rongelapese are unable to cultivate the land where they currently live, they have no means to provide for their iroij. Lacking land and the means to produce customary tributes, the Rongelap community now issues $5,000 quarterly payments to the iroij of Rongelap as tribute. The Rongelap local government also pays $3,000 per quarter to the Mejatto *alab*. Compensatory damages need to be structured in ways that respect the reciprocal exchanges, obligations, and responsibilities of and between each of the three tiers of Marshallese society.

Youth Concerns

The Rongelapese have expressed deep concern about the fate of Rongelap's youth. Youths have not had the same breadth of experience with or connection to their property on the outer islands that their elders have had. Furthermore, Rongelapese adults worry that youths will be unable to exercise or understand their full range of property rights in the future:

> **Johnsay Riklon**: *Kids from Rongelap that live in Majuro are not involved in the Rongelap community. There aren't many community activities for them to get involved in. I don't think they feel Rongelapese. They don't know their relatives. Their friends are kids from other atoll*

---

[20] Mike Kabua, advisory committee meeting, Majuro, March 2, 1999.

communities. I doubt kids will want to go back to Rongelap. It will be a big adaptation from Coca-Cola to coconuts. Youth are not involved much in voting or other activities.[21]

George Anjain: *When you leave home, there is no role for the kids. They used to work and contribute, but not anymore.*[22]

Kajim Abija: *My kids are Rongelapese, but they are from Kwajalein.*[23]

Jerkan Jenwor: *When I was a youth on Rongelap, I started to fish when I was about eight years old. I knew how to make copra and do all kinds of work. Eight to twelve years of age are important learning years.*[24]

Rongelapese youths also have concerns. Informants voiced worries about decisions the elderly might make regarding future resettlement of Rongelap. Youths fear that elders haven't had much education and might not make the right decision about resettlement. Rongelapese youths also worry about the range of opportunities available to them in the urban and displaced areas:

Ken Kedi: *Rongelapese youth can't climb trees, but they are familiar with Coca-Cola. Youth used to keep busy and fit doing work in their environment, such as making copra. They can't do that in the urban areas, however, and they are unfit as a result.*[25]

Compensatory strategies for damages to the people and environment of Rongelap, Rongerik, and Ailinginae should acknowledge the rights and needs of Rongelap youth and future generations to survive from their property and resources. Youths also need sports equipment, education, and opportunities to learn about the unique history and survival techniques of their elders and ancestors.

Needs of the Elderly

The older Rongelap community members suffer greatly from the testing program. They were exposed to near-lethal amounts of radioactive fallout; they suffer from a range of health problems; they get ping-ponged between medical programs without adequate attention; they watch friends and children die; they were relocated many times; they are displaced from their land; they are anxious about death away from their homeland and proper burial grounds; they are unable to obtain their basic

---

[21] Johnsay Riklon, interview by Holly M. Barker, Barbara Rose Johnston, and Stuart Kirsch, Majuro, February 28, 1999.
[22] George Anjain, advisory committee meeting, Majuro, March 2, 1999.
[23] Kajim Abija, interview by Holly M. Barker, Ebeye, March 16, 1999.
[24] Jerkan Jenwor, interview by Holly M. Barker, Majuro, March 17, 1999.
[25] Ken Kedi, interview by Holly M. Barker, Barbara Rose Johnston, and Stuart Kirsch, Majuro, March 1, 1999.

needs; and they worry about the future of their children and the community. Furthermore, the elders lack the traditional respect they would receive on their own land based on property ownership and knowledge about survival. Younger Rongelapese lack the monetary resources needed to adequately provide for the elderly as they did on Rongelap, Rongerik, and Ailinginae. As a result, elderly Rongelapese are sick, often hungry, and dependent on others for their needs. In their displacement, their ties to their land, knowledge about survival, and land rights are less and less relevant to daily life. The elderly are concerned that if they die, no one will pass on information about their property and about their experiences during the testing to youth.

James Matayoshi: *The elders mostly want to return to Rongelap because they know no other way of life.*[26]

Compensation for Rongelap, Rongerik, and Ailinginae must include ways for the elderly to provide for themselves to make up for the loss of land, critical resources, and the family's traditional ability to care for the elderly. Some informants suggested that rather than providing outside assistance to support the individual needs of the elderly, compensation schemes should support families in ways that allow them to fulfill their obligations to the elderly.

Compensation Issues of Women

Land is passed down matrilineally in the Marshall Islands. Because the Rongelapese no longer live on their land and are unable to use their contaminated property, the authority that women possess as inheritors of the land is attenuated. Displacement also means that Rongelapese women have lost their ability to make handicrafts and household supplies, because on Ebeye and Majuro, they don't have access to the materials they need. As a result, Rongelapese women do not have the ability to generate an income as they did on their own property:

Timako Kolnij: *On Rongelap, young girls like me learned to make preserved pandanus and food from pandanus. We cooked it in the underground oven. We also made sitting mats, sleeping mats, and baskets. Our grandmothers taught us. We prepared shells for handicrafts. We got the pandanus leaves ready for the women. We got the pandanus leaves from the middle of the island. We didn't ask if we could cut the pandanus. We all lived together from the resources. I also prepared a lot of ripe breadfruit.*[27]

---

[26] James Matayoshi, interview by Holly M. Barker, Barbara Rose Johnston, and Stuart Kirsch, Majuro, March 1, 1999.
[27] Timako Kolnij, interview by Holly M. Barker, Majuro, March 18, 1999.

Compensation schemes must be careful not to place monetary compensation for a fixed period of time solely in the hands of the Marshallese men who manage the land for the women who carry the rights to the land. It is important to recognize and not undermine the power and position in society that women retain as a result of their land stewardship. Furthermore, compensation for the loss of use of Rongelap, Rongerik, and Ailinginae should include means for women to participate in the economy. Training and education will be required to help Rongelapese women find new ways to contribute to family incomes.

Compensation Issues of Men

Rongelapese men are proud of the fishing, sailing, and navigation skills they culti-vated on Rongelap, Ailinginae, and Rongerik. On their homelands, Rongelapese men had a range of fishing and food-gathering techniques that ensured their ability to provide food for their families, as is expected of Marshallese men. In the areas where the Rongelapese now live, the ability to survive depends largely on the ability to generate a cash income. Rongelapese men believe they lack the educational and professional backgrounds required to obtain wage-earning jobs. The interviewees note that suicide has increased dramatically among young Marshallese men. They believe the high rates are due to a perceived loss of worth.

Ken Kedi: *We have lost our knowledge, our ability, our moral standing and self-esteem in the community. What we were taught is no longer practical. To be a good fisherman, you have to know where to fish on an island. A lot has been lost, not just our land.*[28]

George Anjain: *It is really hard for the men to practice fishing skills they know so well.*[29]

Compensation schemes should consider the long-term ability and qualifications of Rongelapese men to provide for their families. If the Rongelapese cannot rein-habit their atolls because of lingering contamination, compensation should consider the training and employment needs of Rongelapese men.

Research Needs

Research Principles

Although the Rongelapese express anger and resentment about being the subjects of numerous research experiments, they recognize that investigations must still

---

[28] Ken Kedi, interview by Holly M. Barker, Barbara Rose Johnston, and Stuart Kirsch, Majuro, March 1, 1999.
[29] George Anjain, advisory committee meeting, Majuro, March 2, 1999.

be made into areas that are not understood. Informants provided a number of suggestions to guide future research:

- Research in the Marshall Islands should be conducted in ways that assist the Marshallese in one area or another.
- The goal of research on living subjects should be to improve their quality of life.
- All research should be conducted in a transparent manner, with community participation in every aspect of the research and reporting.
- Epidemiological investigation and health care should be provided for the full range of radiation-related problems that the Rongelapese experience—including reproductive abnormalities and problems experienced by the second and third generations—and not just the illnesses that outsiders want to consider.
- Medical research and health care should be based on a wide range of individual experiences rather than on generalizations about control groups compared to "exposed" groups.
- All research must be translated into Marshallese. Results of research must be discussed with and turned over to the community.

Research Questions

Specific research questions that the Rongelapese want to see addressed include:

- Is it really safe to return to Rongelap?
- If some areas are too contaminated for habitation, can they be safely used for short periods of time?
- If some foods, such as the coconut crab, are too contaminated for regular subsistence use, can they still be consumed on special occasions?
- Many species of fish that did not cause fish poisoning before the nuclear tests became poisonous after the tests. Is there a correlation between fish poisoning (from ciguatera or other sources) and nuclear testing?
- Is it safe to consume reef fish, shellfish, and other marine resources from the northern islands?
- Is it safe to drink water from wells or water stored in catchments or cisterns made from local materials?
- If lagoon sediments are so heavily contaminated with plutonium, is it safe to swim and bathe in lagoon waters?
- Are there times when women (because of menstruation, childbirth, and other reproductive tract changes) increase their vulnerability by swimming and bathing in plutonium-contaminated lagoon waters or squatting close to contaminated soil?

- Can hot spots and contaminants be monitored to let us know where it is safe?
- Environmental surveys to date have sampled for the presence of a few radio-isotopes. A holistic approach to environmental contaminants in soil, water, and vegetation is lacking. Where are all the environmentally hazardous sites? In addition to hot spots from radioactive fallout and bioaccumulation, where are the sites where naval ships were decontaminated and scuttled, and sites where toxics such as PCPs were disposed? Tracer chemicals include a range of persistent radioactive isotopes and other toxins, such as arsenic. Are there current and future environmental risks from exposure to persistent tracer chemicals? Are there any other environmental hazards associated with U.S. military use in the northern islands?
- Is it safe for women to get pregnant and raise children in the northern islands? What are the mutagenic effects of long-term systemic exposure?
- Is it safe for people who were acutely exposed in the past to return to a setting where they will receive constant low-level exposure? Do assurances of safety consider the effects of contaminants on individuals whose health has already been compromised by long-term cumulative exposures? Do current protection standards protect people who were previously exposed?
- Does exposure to radionuclides contribute in some way to increased susceptibility to other ailments (synergistic effects)?
- How can we reduce risks by modifying behavior? How do different factors such as eating habits, exposure to smoke from cooking fires, and different work activities increase or reduce risks from radionuclides and other environmental contaminants? Are there different exposure pathways and risks for women, men, children, teenagers, and the elderly? Are there ways to reduce intergenerational risks? What behavioral changes are needed to reduce risks?
- How does radiation affect the overall health of families (not just individuals)?

Ideas for Remedial Action

Community Health Care Needs

Suggestions for improving community health include:

- Medical centers for all members of the Rongelap community, not just the small number of people eligible to participate in health programs defined in the Compact of Free Association.
- A women's health education and outreach program, with mobile clinics staffed by Marshallese women trained to give maternity and gynecological exams, treat infections, and do breast and reproductive-tract cancer screening procedures.
- Counseling services to assist the Rongelap community with domestic problems, depression, and substance abuse.

- Establishment of cancer and thyroid registries for the Rongelap community to better understand disease incidence levels and to give the Marshallese scientific and health community the means to generate reliable statistics on the health consequences of the nuclear testing program.

Social and Cultural Health

Community health should also reflect the social and cultural health of the community. Suggestions in this area include:

- **A Rongelap Community Center** A community center would be a central meeting place for those scattered throughout the Marshall Islands and for those living in the immediate area. The Rongelapese envision a large building constructed with local materials. Construction of the building would be part of the center's educational program, as workers would share their knowledge of traditional building materials and design. The center would serve as a meeting place where older Rongelapese could teach the younger generation about their history and land rights and the traditional knowledge and skills necessary for survival. Rongelapese elders could instruct youths in fishing techniques, boat building, weaving, food cultivation, cooking, navigation, chants, dancing, and so forth. Food prepared by the center could be distributed to disabled and elderly Rongelapese. Some members of the community would look after small children so that more adults could participate in the center's activities. Lighting would enable the center to operate in the evenings, when it is cooler and people are not working. Solar power and computer terminals would enable the center to communicate with Rongelapese community members scattered throughout the RMI and the United States. The center could collaborate with the larger community, local schools, and the College of the Marshall Islands in various educational and outreach programs.
- **Job Training** When the Rongelapese lived on their own lands, they had the skills and knowledge needed to provide for their families. In their displacement, the Rongelapese do not have the educational background or job skills necessary to obtain the wage-earning jobs they need to support themselves. Special job training programs should be made available to prepare the Rongelapese for employment in growing sectors of the RMI's economy, such as fisheries and aquaculture, tourism, and construction. Job training should capitalize on customary knowledge, techniques, methods, and materials.
- **National Treasures** Similar to the designation of important historic locations as historic trusts, Rongelapese people who are recognized masters in their command of history, customary knowledge, or traditional skills should be designated national treasures. Master navigators, builders, dancers, singers, fishers, weavers,

healers, and so forth should be honored with a title and lifetime salary. They should be bestowed with the responsibility of passing on their knowledge to the younger generation in meaningful ways, such as instruction in the local schools and community center and by sharing their pictures, stories, records, and skills with the community.

- **Oral History** The number of Rongelapese who were alive during the U.S. nuclear weapons testing program is dwindling rapidly because of illness and age. It is imperative to interview survivors of the Bravo test and those who understand the history of the community. The oral histories of all Rongelapese alive during the testing program should be collected and archived. Ideally, Rongelapese youths should be trained in oral history collection techniques so they can be active participants in gathering information and learning their history from their elders. The oral histories should be videotaped, transcribed, and translated. Effort should be made to capture the memories and experiences in situ—with stories and testimony documented on Rongelap, Ailinginae, and Rongerik. This process will enable the community to preserve important source material about its history and homeland. The oral histories will educate subsequent generations of Rongelapese as well as the larger public about the experiences of the Rongelapese.

- **Higher Education** Scholarships should be set aside for Rongelapese students to pursue collegiate and graduate studies in subjects that are important to the health and well-being of the community, including its economic future. Scholarships should support training in radiation science, medicine, marine biology, toxicology, environmental health, natural resource management, tourism, education, and other academic and trade priorities defined by the community.

- **Museum** A museum should be erected on Rongelap or another location to document and display the Rongelapese people's history and experiences with the U.S. nuclear weapons testing program. Rongelapese who remember the events surrounding the Bravo test could give tours of the museum to patrons. The museum would educate younger Rongelapese and the Marshallese public. Furthermore, the museum would attract international visitors and provide economic assistance to the community, particularly if the community decides to return to Rongelap.

Community Infrastructure

- **Safe Food** Rongelap informants were especially concerned about food safety. Because of concern about residual radiation and lack of space, some Rongelap informants suggested building a community greenhouse using hydroponics to grow fruits and vegetables in a safe environment. A greenhouse would provide nutritious foods to the Rongelap community. It would also provide educational and employment opportunities for the community. An alternative agricultural research facility might look at food-production techniques for use in the Marshall Islands. For example, lagoons might serve as viable locations for floating gardens and allow the production of food for land-scarce communities such as Ebeye.

- **Realistic Transportation** Transportation is a major problem that impedes the ability of the Rongelapese to come together and function as a community. Some informants suggested supplying all islands with free bicycles for general transportation use. Bicycles would provide a means of exercise and transportation to schools, the community center, jobs, and homes on the islands where community members reside. Ferries or boats are also needed to move resources and people between islands, especially to and from Mejatto. Seaplanes or helicopters are needed to provide transportation for medical and other emergencies on islands with no medical facilities.

- **Information Access and Communication** Like transportation, communication is a major problem, especially when the community lives in such widely separated spaces. Community integrity would be significantly enhanced with the means to regularly communicate. Some informants discussed the need for nationwide access to computers, modems, satellite hookups, and independent power sources (such as wind generators and solar battery packs). Computers could be made available at public schools, community centers, and other public settings on all residential islands.

Environmental Concerns

Environmental concerns voiced by informants included the inability to maintain a stewardship presence on their land; the inability to protect fish, clams, birds, turtles, and other valuable resources from outside exploitation (especially foreign fishermen stripping unmonitored resources); the difficulty of restoring Rongerik's devastated ecosystem; the inability to utilize plants, water, and marine resources due to contamination; and the biodegenerative conditions of valued resources.

Recommendations included developing a comprehensive plan for environmental assessment, monitoring, remediation, and restoration; implementing plans in ways that develop Marshallese capacity to effectively address the natural resource management needs of Rongelap, Rongerik, and Ailinginae; and integrating natural resource restoration and management work with economic development strategies such as the promotion of ecotourism.

Part 5

Conclusions and Recommendations

Violations of Trustee Relationships

The United States exerted territorial authority over the Marshall Islands by establishing a military presence there during World War II. U.S. presence and authority significantly increased with the onset of the nuclear weapons testing program in 1946. In 1947 the United Nations formally recognized this territorial relationship, designating the Marshall Islands as a trust territory of the United States.

International trusteeships, established by United Nations Charter, were meant to "promote the political, economic, social, and educational advancement of the inhabitants of the trust territories, and their progressive development towards self government or independence."[1] Trustee relationships between state and domestic dependent nations have been acknowledged and confirmed in constitutions, treaties, and enabling legislation. Trust relationships represent an enforceable legal acknowledgement by a state that it has taken what once belonged to trustee (native, indigenous, aboriginal) peoples and that it agrees to protect what they retain.[2] Trust relationships involve a fiduciary relationship, in which the state is legally bound to "act for the benefit of the other while subordinating one's personal interest."[3]

In addition to the rights and responsibilities outlined in trusteeship agreements, inhabitants of trust territories are "protected peoples," whose status and conditions are protected by rights outlined in UN covenants, including the Geneva Conventions

---

[1] Charter of the United Nations, articles 75 and 76 (June 26, 1945).
[2] See Mary Christina Woods, "Fulfilling the Executive's Trust Responsibility Toward the Native Nations on Environmental Issues: A Partial Critique of the Clinton Administration's Promises and Performances," *Environmental Law Journal* 25 (1995):733, 742, discussed in Hyun S. Lee, "Post Trusteeship Environmental Accountability: Case of PCB Contamination on the Marshall Islands," *Denver Journal of International Law and Policy* 26, no. 3 (Spring 1998): 424–25.
[3] *Black's Law Dictionary,* 6th ed. (St. Paul, MN: West Publishing, 1990), 626.

of 1949 (defining medical experiments on protected persons as a grave breach and crime against humanity) and the 1998 Rome Statute of the International Criminal Court (identifying medical experiments as war crimes, whether they occur in an international or internal context). The Rome Statute defines as a crime: "Subjecting persons who are in the power of an adverse party to the physical mutilation or to medical or scientific experiments of any kind which are neither justified by the medical, dental or hospital treatment of the person concerned nor carried out in his interest, and which cause death to or seriously endanger the health of such person or persons."

Under U.S. law, inhabitants of a trust can sue to enforce their treaty rights. There are literally thousands of examples of this type of suit filed by Native American tribes seeking remedy for damages from violations of treaty rights. Similarly, the International Court of Justice has ruled that inhabitants of an international trusteeship can sue for violations of substantive rights and duties established in trusteeship agreements.[4]

This report includes documentation of various violations of trusteeship responsibilities as a result of the U.S. nuclear weapons testing program:

- Atmospheric testing in the Marshall Islands (1946–1958) demonstrated military might but also inflicted nuclear war conditions on a fragile atoll ecosystem and a vulnerable population.
- Devastation and contamination was documented from the onset of testing, and despite requests from the Marshallese in 1954 to cease using their atolls as proving grounds, the United States continued to detonate nuclear weapons, which resulted in additional fallout on Rongelap and other populated atolls.
- Nuclear weapons testing destroyed the physical means to sustain and reproduce a self-sufficient way of life for peoples living in the northern atolls and produced great hardships for the nation as a whole.
- Radioactive contamination and involuntary relocation radically altered health, subsistence strategies, sociopolitical organization, and community integrity.
- Human subject research included the willful return of an exposed population to a known hazardous setting for the purpose of documenting the long-term and cumulative effects of continual exposures to nuclear weapons fallout.
- Human subject research experiments violated legal and ethical responsibilities of medical research and informed consent.

---

[4] *Nauru v. Australia*, International Court of Justice 240, June 26, 1992. During its trusteeship period, Australia mined phosphate, removing approximately one-third of the island and leaving the remainder in a degraded state. The International Court of Justice ruled that it had jurisdiction to hear the case, but Australia and the Republic of Nauru settled their claims before the ICJ could issue a ruling. Australia agreed to pay Nauru $107 million (Australian) to facilitate post-phosphate economic development.

- Human subject research activities included experiments conducted without meaningful consent for purposes that had no direct benefit to individuals.
- Human subject research experiments included multiple incidents of additional exposures to radioactive materials producing pain and suffering and further endangered the health of human subjects by increasing cumulative doses.
- Responding to the medical needs of an exposed population by an intensive research program, rather than a holistic health treatment program, produced otherwise avoidable incidents of pain and suffering.

Statements of Culpability

In addition to the careful documentation of damage and injury contained in the recently declassified medical surveys and lab reports cited in this report, culpability for radiation-related injuries has been publicly acknowledged in U.S. statements to the United Nations and in treaties and agreements between the Republic of the Marshall Islands and the U.S. government.

The April 20, 1954, petition submitted to the United Nations Trusteeship Council on behalf of the Marshallese citizens of the Trust Territory of the Pacific Islands cites Marshallese concerns about the U.S. nuclear weapons testing program. Concerns included damage to health and the long-term implications of being removed from their land:

> We, the Marshallese people feel that we must follow the dictates of our consciences to bring forth this urgent plea to the United Nations, which has pledged itself to safeguard the life, liberty and the general well being of the people of the Trust Territory, of which the Marshallese people are a part.
>
> . . . The Marshallese people are not only fearful of the danger to their persons from these deadly weapons in case of another miscalculation, but they are also very concerned for the increasing number of people who are being removed from their land.
>
> . . . Land means a great deal to the Marshallese. It means more than just a place where you can plant your food crops and build your houses; or a place where you can bury your dead. It is the very life of the people. Take away their land and their spirits go also.[5]

The United Nations Trusteeship Council's response to the Marshallese petition noted, "The Administering Authority adds that any Marshallese citizens who are removed as a result of test activities will be reestablished in their original habitat in such a way that no financial loss would be involved."[6]

---

[5] Petition from the Marshallese People Concerning the Pacific Islands: Complaint Regarding Explosions of Lethal Weapons within Our Home Islands, to United Nations Trusteeship Council, April 20, 1954.

[6] United Nations Trusteeship Council, Petitions Concerning the Trust Territory of the Pacific Islands, July 14, 1954, 5.

In response to this petition and the resulting international inquiry at the United Nations, the United States assured the Marshallese and the nations of the world that:

> *The fact that anyone was injured by recent nuclear tests in the Pacific has caused the American people genuine and deep regret. . . . The United States Government considers the resulting petition of the Marshall Islanders to be both reasonable and helpful. . . . The Trusteeship Agreement of 1947 which covers the Marshall Islands was predicated upon the fact that the United Nations clearly approved these islands as a strategic area in which atomic tests had already been held. Hence, from the onset, it was clear that the right to close areas for security reasons anticipated closing them for atomic tests, and the United Nations was so notified; such tests were conducted in 1948, 1951, 1952 as well as in 1954. . . . The question is whether the United States authorities in charge have exercised due precaution in looking after the safety and welfare of the Islanders involved. That is the essence of their petition and it is entirely justified. In reply, it can be categorically stated that no stone will be left unturned to safeguard the present and future well-being of the Islanders.[7]*

In a related press release issued by the United States Mission to the United Nations, "Guarantees are given the Marshallese for fair and just compensation for losses of all sorts."[8]

Statements of U.S. culpability are also found in the 1983 Compact of Free Association (approved by the U.S. Congress in 1986). Section 177 of the Compact of Free Association states:

> *The Government of the United States accepts the responsibility for compensation owing to the citizens of the Marshall Islands, or the Federated States of Micronesia (or Palau) for loss or damage to property and person of the citizens of the Marshall Islands, or the Federated States of Micronesia, resulting from the nuclear testing program which the Government of the United States conducted in the Northern Marshall Islands between June 30, 1946 and August 18, 1958.[9]*

Section 177 of the agreement outlines U.S. responsibility for the consequences of the nuclear weapons testing program and provides a one-time payment of $150 million to the Republic of the Marshall Islands to create a trust fund for addressing

---

[7] Mason Sears, U.S. representative, statement to the Trusteeship Council, United States Mission to the United Nations. Press release 1932, July 7, 1954, http://worf.eh.doe.gov/data/ihp1d/400107e.pdf (accessed October 25, 2007).

[8] Frank E. Midkiff, high commissioner of the Trust Territory of the Pacific Islands, statement to the United States Mission to the United Nations. Press release 1932, July 7, 1954, http://worf.eh.doe.gov/data/ihp1d/400107e.pdf (accessed October 25, 2007).

[9] Compact of Free Association Act of 1985, Public Law 990239.

past, present, and future claims arising from the testing program. Included in the compact is a "changed circumstance" clause that allows the RMI to petition Congress for additional funding and assistance, if the RMI can (1) demonstrate the existence of new and additional information about the effects of the testing program; (2) demonstrate that this new information was not known during compact negotiations; and (3) based on this new information show that U.S. compensation provided in the compact is manifestly inadequate.[10]

The United States has also publicly acknowledged the occurrence of non-therapeutic biomedical research involving the administration of radioisotopes to the Rongelapese and other Marshallese subjects. Furthermore, this biomedical research program occurred with "some tension between data gathering and patient care"; "the additional strains of language and cultural differences between the Marshall Islanders and the physicians appears to have compromised the process of informing the subjects of the purposes of the tests and of obtaining their consent." Experiments "took place at a time [the mid-1950s] when the government rules requiring disclosure and consent in the use of radioisotopes with healthy subjects were established and public; the available documented evidence suggests that these rules were not followed."[11]

Following the published results of the ACHRE review of documents and testimony, the Rongelap Atoll local government asked the DOE to review patient records and identify Rongelapese participants in an anemia experiment referred to by ACHRE and involving Cr-51 and tritiated water. In April 1999, the DOE disclosed to the Rongelap local government its ability to identify all twenty-one participants who received radioisotopes in the 1963 BNL anemia studies and "less than complete information" for the 1961 and 1962 studies involving thirteen Marshallese participants, including six patients used in at least two studies.[12]

Reparations

In the years since the initial compact was signed, thousands of documents, demonstrating a much broader area of contamination and greater numbers of

---

[10] Ibid.

[11] Advisory Committee on Human Radiation Experiments, *Final Report,* 585.

[12] See Paul Seligman, letter to Mayor James Matayoshi, April 29, 1999, in the appendix of this report. While this letter acknowledges culpability in one incident of Cr-51 exposure, documents subsequently released through the Clinton administration declassification order demonstrate that exposures to Cr-51 and other radioisotopes were not isolated incidents. For example, the 1959 medical survey includes references to the use of Cr-51 to tag plasma volume in "10 exposed and 10 nonexposed people." Lab reports sent to Dr. Conard report Marshall Islands blood-volume data derived from Cr-51 from twenty-one Rongelapese samples in 1964 (http://worf.eh.doe.gov/data/ihp2/4625_.pdf) and eighteen Rongelapese samples in 1965 (http://worf.eh.doe.gov/data/ihp2/4609_.pdf).

affected people than previously acknowledged, have been declassified. Also, since the Compact of Free Association was first crafted, scientific research has further demonstrated a wide range of health risks associated with low-level radiation exposure, including the cumulative effects of low-level exposure over time. This report has presented evidence drawn from new information demonstrating a broad range of physical hardships and sociocultural, economic, political, and psychosocial damages experienced by individuals, the community, and the nation. This evidence suggests that U.S. efforts to address its breach of obligation, by providing medical treatment, environmental cleanup, and other remedial actions, were not only insufficient but in a number of instances produced further harm.

Broader levels of compensatory actions are needed to address the much greater injury to people and the environment suggested and substantiated by this new information. One strategy for broadening the approach to compensation is to consider compensatory payments as one of a wide range of remedial actions that provide remedy, or reparations, for damages.[13]

Reparation is defined as action or processes that repair, make amends, or compensate for damages. In a legal sense, there are three generally recognized forms of reparation: restitution, indemnity (or compensation), and satisfaction. The term *reparation* generally refers to remedial actions meant to repair a breach of obligations established in international law.

Restitution is designed to put the offended state back in the position it would have been in had the breach not occurred. Restitution may include performance of the obligation, revocation of the offending act, or abstention of the unlawful conduct. Restitution in this case should include U.S. actions that seek to restore a healthy ecosystem and sustainable way of life, including efforts to mitigate adverse socioenvironmental impacts by reducing individual and household risk, improving degraded marine and terrestrial ecosystems, increasing awareness of and involvement in risk-management activities, and providing education and training opportunities that ensure Rongelapese control over the knowledge and health of their environs.

Indemnity, also termed compensation, involves the payment of money to the offended party for any losses incurred by the illegal act, including any lost profit or value of lost property. As used by the Republic of the Marshall Islands Nuclear Claims Tribunal, the term *compensation* refers to indemnity payments to affected peoples to compensate for bodily injuries and the loss of assets and property. Indemnity payments as part of reparations agreements have been used to provide individual

---

[13] This discussion of reparations and the case summaries of precedents for compensation for loss of a way of life experienced by Native Americans and awarded by U.S. courts are drawn from Barbara Rose Johnston, *Reparations and the Right to Remedy* (World Commission on Dams, 2000), http://www.dams.org/docs/kbase/contrib/soc221.pdf (accessed April 1, 2008).

cash awards as well as to fund a variety of remedial actions, including resettlement plans and development programs. However, in many cases compensation has been granted only to those who can demonstrate (1) legal ownership and (2) individual claim with loss value calculated according to prevailing market rates as an average of registered sales prices (of land and other economic assets) in the recent past. Thus this form of reparation generally represents indemnity payments based on market values rather than replacement values.

Satisfaction includes almost every other form of reparation and is meant to address any nonmaterial damage. Examples of satisfaction include public acknowledgements that a wrong was committed, formal apologies, and discipline of guilty individuals.[14] Satisfaction also may include damage awards for hardships encountered as a result of long-term and cumulative effects of the original breach of obligation.

Relevant Case Precedents

Until recently, the majority of cases involving reparations acknowledged and compensated victims of war-related atrocities, especially those surrounding the events of World War II. These cases include acknowledgement of and efforts to seek some sort of remedy for injuries suffered by protected persons as unwilling subjects in medical experiments. With the creation of post–World War II human rights treaties and the expansion of international and national human rights and environmental law, a broader range of rights has been acknowledged and a broader range of abuses or violations of rights has been documented. Increasingly, reparations are being made to redress violations of international and domestic law committed in the name of colonial expansion and related violations of trustee relationships, economic development, and national security. Examples of restitution, indemnity, and satisfaction relevant to this hardship and consequential damages claim are briefly summarized below.

In 1948 the U.S. Congress passed the Japanese American Evaluation Claims Act, which provided funds to pay claims for real and personal property losses to the 120,000 people of Japanese ancestry who were forced to abandon their property and reside in federal internment camps during World War II. Some $38 million has been paid under this act. In 1988, as part of the Civil Liberties Act, the U.S. Congress established a $1.65 billion restitution program for persons of Japanese

---

[14] Examples of public apology as reparation include written apologies issued by the U.S. government to Japanese families imprisoned in U.S. internment camps during World War II and apologies from President Clinton to the indigenous people of Hawaii for unlawful seizure of the lands and resources of a sovereign nation by past administrations of the U.S. government. Examples of criminal investigations and prosecution include the February 25, 2000, sentencing of the former director of the District Construction Bureau in Fengdu, China, who received the death sentence for stealing 12 million yuan (1.44 million dollars) from Three Gorges Project accounts.

ancestry who were forced to relocate to internment camps. The program author-ized $20,000 reparation payments for hardship and indignity experienced by some sixty thousand eligible people.

The 1988 Civil Liberties Act also provided some $12,000 each to 450 surviving Aleutian Islanders who were removed from their homeland by the U.S. Navy in 1942 and relocated to abandoned canneries and mines in southern Alaska for three years. An additional $1.4 million was authorized to establish a trust fund to be used for health, education, cultural preservation, community development, and other projects meant to improve the condition of Aleut life. Another $15 million was allocated to compensate for the loss of Attu Island, used as a Coast Guard station during and after the war and designated as a wilderness area in 1980.

Loss of Critical Resources and Damages to a Way of Life

Examples of restitution, indemnity, and satisfaction for the consequential damages accompanying violations of trusteeship responsibilities are also found in Native American case law. These cases often produce reparations awards for violations of trusteeship responsibilities that result in loss of land and other critical resources, producing significant damage to social structures, cultural institutions, and cus-tomary ways of life. In many cases, remedial actions for "loss of a way of life" have included compensation for the loss of the means of subsistence, efforts to restore damaged human and environmental systems, and efforts to provide the means to develop community capacity to regain meaningful and self-sufficient ways of life. These actions can incorporate a range of activities, including appropriations, trust funds, program assistance, and other forms of technical assistance. These con-siderations are present in the Zuni land claims case outlined below.

The aboriginal territory that supported the Zuni way of life originally stretched from the Grand Canyon to the Rio Grande in central New Mexico. Today, the Zuni control some 750 square miles of trust lands, the bulk of which are located in New Mexico, with a smaller, noncontiguous section in Arizona. To reclaim treaty-protected lost lands, the Pueblo of Zuni filed a series of lawsuits in federal courts. In a case filed in U.S. claims court on May 27, 1987, the Zuni tribe of New Mexico claimed that their land was taken without compensation when the United States acquired the land from Mexico. Oral histories with Zuni elders provided initial documentation of traditional land use, and this information was corroborated by scientific and historical documents. Contested issues included whether the Zuni claim to a large area of land and the resources contained within could be substantiated lacking individual title and based on a record of subsistence-oriented use, and whether damages could be awarded for injuries and loss of a sustainable a way of life. The tribe used historic records and informant testimony to "demonstrate how Zuni land use involved a core area of permanent and seasonal settlements, where agriculture was intensively practiced, surrounded by a larger sustaining area . . . with extensive grazing, hunting, and gathering of numerous plants and minerals." The court ruled in favor of the Zuni, finding that lands were taken by the U.S. government without adequate compensation.[15] The ultimate finding of the court was that while the Zuni resided in specific locations, "This entire claim area was used by the Zuni for one purpose or another including: habitation . . . and life-sustaining activities including farming, hunting, grazing, gathering, and religious worship."[16]

The result of the court's finding was the Zuni Land Conservation Act of 1990 (P.L.101-486). Under this act, the Zuni and the U.S. Department of Interior were instructed to formulate a resource plan including:

- A methodology for sustained development of renewable resources
- A program of watershed rehabilitation
- A computerized system of resource management and monitoring
- Programs for funding and training Zuni Indians to fill professional positions that implemented the overall plan
- Proposals for cooperative programs with the Bureau of Indian Affairs and other private or public agencies to provide technical assistance in carrying out the plan
- Identification and acquisition of lands necessary to sustain Zuni resource development.

Congress established a $25 million trust fund to formulate and implement the plan. Portions of the trust are set aside to pay outstanding debts of the tribe, construct a public elementary school and increase educational opportunities, and purchase land for the community. It was recognized that this action is essential to "the preservation of the Zuni history, culture, tradition, and religion."[17]

Findings in the Zuni cases reinforce earlier court rulings on Native American claims that recognized that compensation for material losses experienced by individual property owners does not adequately encompass the corporate losses experienced by a group whose way of life revolves around subsistence-oriented use of natural resources. For example, when the southern California Soboba band of Mission Indians went to court to claim compensation for damages from lost water resources, economist Raul Fernandez used ethnographic material compiled by

---

[15] Stephen G. Boyden, "The Zuni Claims Cases," In *Zuni and the Courts: A Struggle for Sovereign Land Rights,* ed. E. Richard Hart (Lawrence: University Press of Kansas, 1994), 225.
[16] E. Richard Hart, ed. *Zuni and the Courts: A Struggle for Sovereign Land Rights* (Lawrence: University Press of Kansas, 1994), 245–77.
[17] Boyden, "Zuni Claims Cases," 225.

anthropologist Joe Jorgensen to demonstrate that impoverishment and sociocultural disintegration were linked to the building of the Colorado River aqueduct, which caused tribal water resources to dry up and eventually disappear. Compensation had been initially offered to individual landowners. The tribe contested this solution as inadequate. The court found that the tribe as a whole had experienced damage from the loss of "natural capital" represented by naturally occurring water resources on their reservation. Court rulings acknowledged that compensation to individuals for the loss of water and agricultural production did not adequately compensate for broader sociocultural losses. Thus court rulings included mandates to restore water, and restitution granted to the entire tribe to support social reconstruction of the Soboba community.[18]

Other relevant Native American case precedents include awards for consequential damages from actions that damage or destroy subsistence and commercial fisheries. For example, construction of the Grand Coulee Dam in 1933 displaced an estimated 5,000 people, including 1,500 Colville Indians and 100 to 250 Spokane Indians. Neighboring Nez Percé, Coeur d'Alene, Warm Springs, and Yakama tribes were also affected. Displaced peoples received inadequate compensation for land and goods, and no financial support or assistance for resettlement. Many inhabitants of submerged towns were relocated to areas lacking basic utilities, such as a water source, public telephones, and electricity. Some rejected the government's compensation offers as being too low. The government responded by condemning their land to obtain the titles. Subsequent efforts in court to secure just compensation were not successful. Native Americans objected to failures to live up to compensatory promises, such as Interior Secretary Harold Ickes's promise that the tribes would receive a share of hydroelectric power revenues. Dam construction hampered intertribal communication, resulted in the loss of sacred sites and cemeteries, and caused a significant decline in native fish populations, including five species of salmon and two species of trout. The loss of fish resulted in the loss of the tribal way of life.

---

[18] R. Fernandez, "Evaluating the Loss of Kinship Structures: A Case Study of North American Indians," *Human Organization* 46 (1987): 1–9. See also J. Stephen Lansing, Philip S. Lansing, and Juliet S. Erazo, "The Value of a River," *Journal of Political Ecology* 5 (1998): 1–23. The authors apply Fernandez's ideas on natural capital in their assessment of the damages experienced by the Skokomish Indian Reservation in western Washington with the building of the Cushman Dam on the north fork of the Skokomish River in 1930. They argue that the biological productivity of a river ecosystem is an essential component of the "corporate estate" of the Skokomish tribe. Losses to the biological productivity of the river ecosystem can be seen as reductions in immediate income derived from the river as natural capital. Since 1930 each individual member of the tribe has experienced loss of access to diverse riverine resources, and the "costs to the tribe are analogous to losses in capital rather than immediate income." Thus compensation for this loss should include investments in the social and economic infrastructure of the tribe to help restore the depleted value of tribal institutions, in addition to the restoration of the natural systems that formed the basis of the tribe's cultural traditions and "enabled the steady accumulation of natural capital."

In 1951 the Colville Confederated Tribes filed suit against the United States. The Indian Claims Commission divided the suit into two cases. Docket 181-C contained claims for the loss of fisheries and the elimination of salmon-run populations as a result of dam construction. Docket 181-D covered compensation for annual power share revenue from tribal land that had been promised to the tribes. In 1978 the commission ruled on Docket 181-C that the U.S. government was obliged to guarantee tribal fishing rights. The tribes were entitled to reparations for the difference between the value of the fish they were able to catch between 1872 and 1939 and the value of what their normal subsistence catch would have been. This reasoning produced an award of only $3,257,083, which did not include damages. In Docket 181-D, after two federal court rulings in 1990 and 1992,[19] the United States and the tribe negotiated a settlement. The United States provided partial compensation to the tribe for the damages suffered from the dam. The tribe received a $53 million lump sum settlement for previous years, from funds appropriated by Congress in 1944. The act also provided that, thereafter, the Bonneville Power Administration—the hydroelectric producer that benefited from dams placed on the Columbia River—would make annual payments to the tribe of approximately $15 million.

Natural Resource Damages

In addition to case precedents awarding compensation for past experiences involving the loss of critical resources and related damage to a way of life, case precedents relevant to the Rongelap claim include those awarding damages for current and future conditions involving contaminated or destroyed natural resources. Relevant sources of U.S. law and implementing policies framing the assessment of environmental contamination and the valuation of natural resource damage include:

- *Final Guidance for Incorporating Environmental Justice Concerns in EPA's NEPA Compliance Analyses,* released by the U.S. Environmental Protection Agency, Office of Federal Activities, April 1998
- *Guidance for Addressing Environmental Justice under the National Environmental Policy Act (NEPA),* Council on Environmental Quality, March 1998
- *Considering Cumulative Effects Under the National Environmental Policy Act,* Council on Environmental Quality, January 1997
- Executive Order 12898 on Federal Actions to Address Environmental Justice in Minority Populations and Low-Income Populations with Accompanying Memorandum, February 11, 1994

---

[19] *Coville Confederated Tribes v. The United States,* Indian Claims Commission, docket 181-D (20 Ct. Cl 31; 964 F. 2nd 1102).

- The Oil Pollution Act of 1990
- The National Environmental Policy Act of 1969 as amended, 42 U.S.C. 4321–4347, January 1, 1970
- *Siting of Hazardous Waste Landfills and Their Correlation with Racial and Economic Status of Surrounding Communities,* U.S. General Accounting Office, June 1, 1983
- The United States Comprehensive Environmental Response, Compensation, and Liability Act of 1980 (CERCLA).

These statutes and implementing policies define injury to natural resources and establish procedures for the valuation of damages on the direct and indirect costs of restoring, rehabilitating, replacing, and/or acquiring the equivalent of injured natural resources. For example, in CERCLA (also known as Superfund), the natural resources covered include land, fish, wildlife, biota, air, water, groundwater, drinking water supplies, and other government- or privately owned resources. Damages may reflect the value of the services lost to the public between the time of release (the polluting act) and the time the resources, and the services those resources provide, are returned to baseline conditions (conditions that would have existed had the release not occurred). EPA guidelines on environmental justice assessment procedures state, "In considering direct, indirect, and cumulative impacts on natural resources, analysts must identify and assess the patterns and degrees to which affected communities depend on natural resources for their economic base (e.g., tourism and cash crops) as well as the cultural values that the community and/or Indian Tribe may place on a natural resource at risk." Damages are calculated and used as supporting evidence in federal, state, tribal, and civil court cases against polluting parties, with awards used to fund remedial actions (such as environmental restoration programs, individual and community compensation for lost wages and income from fishing, and so forth).

Local, state, and nongovernmental agencies have also used these statutes to force federal agency attention and resources to problems relating to nuclear waste contamination. Suits filed against the DOE have resulted in its current effort to develop "long-term stewardship" plans for all sites where long-lived wastes and contaminants relating to nuclear energy and weapons production and use remain in place. Per a settlement negotiated in 1999, DOE has established a $6.25 million fund to assist citizens' groups and tribes in conducting independent technical and scientific reviews of environmental management activities at DOE sites. DOE's Office of Environmental Management spends more than $6 billion dollars per year to meet its responsibility for cleaning up contamination, wastes, nuclear materials, and contaminated structures resulting from nuclear weapons production at more than one hundred sites in thirty U.S. states. Estimates for total cleanup of these sites range from $150 to $250 billion. Despite U.S. government acknowledgements of culpability in creating severely contaminated radioactive sites in the Marshall Islands, little effort has been made to meaningfully include Marshall Islands sites in the assessment, monitoring, and development of remediation plans, as is the case with all other nuclear weapons testing sites under DOE's management.

U.S. environmental protection statutes and implementing policies identify strategies to determine the damage to biological and human systems, and these methods are regularly employed in a variety of lawsuits involving contamination and natural resource damage. Compensation for environmental damage has been awarded in tens of thousands of "toxic torts" cases, including those dealing with oil spills; chemical contamination of soil, water, and air from oil refineries, gas stations, various manufacturing industries, and military base activities; and mining and the downstream effects from mining. Compensatory awards have acknowledged direct and indirect damages to natural ecosystems, loss of land and damage to land (including the economic, social, and cultural value of land and property contained within), socioeconomic damages resulting from the stigmatization of contamination, and damages associated with loss of a way of life.

Perhaps the most significant case of compensation for natural resource damage involves the 1989 *Exxon Valdez* oil spill in Alaska's Prince William Sound. The 11 million gallon spill polluted some 1,000 miles of shoreline, killing tens of thousands of birds and marine mammals and causing socioeconomic damage to local communities, fishing industries, and native peoples. In October 1991, the district court approved a settlement between the state of Alaska, the U.S. government, and Exxon, resolving various criminal charges against Exxon as well as civil claims brought by the federal and state governments for recovery of natural resource damages resulting from the oil spill. The 1991 settlement had three distinct parts:

1. Criminal Plea Agreement: Exxon was fined $150 million, the largest fine ever imposed for an environmental crime. The court forgave $125 million of that fine in recognition of Exxon's cooperation in cleaning up the spill and paying certain private claims. Of the remaining $25 million, $12 million went to the North American Wetlands Conservation Fund, and $13 million went to the national Victims of Crime Fund.
2. Criminal Restitution: As restitution for the injuries caused to the fish, wildlife, and lands of the spill region, Exxon agreed to pay $100 million. This money was divided evenly between the federal and state governments.
3. Civil Settlement: Exxon agreed to pay $900 million, with annual payments stretched over a ten-year period. The settlement has a provision allowing the governments to make a claim for up to an additional $100 million to restore

resources that suffered a substantial loss, whose nature could not have been anticipated from data available at the time of the settlement.[20]

Other fines and settlement agreements had brought Exxon's total liability to $3.5 billion by 1994. In 1994 an Alaskan jury awarded $5.3 billion in punitive damages in a class-action suit representing twenty thousand commercial fishermen, American Indians, and others harmed by the natural resource damage caused by the oil spill. Exxon appealed the jury award in the Ninth U.S. Circuit Court. In August 1998, Exxon's request for a new trial on the grounds of jury tampering was rejected. Exxon appealed this ruling, and on October 2, 2000, the U.S. Supreme Court upheld the *Exxon Valdez* punitive damage award.[21]

Stigma Damages

Stigma damage is a reduction in the value of property caused by increased risk associated with the historical presence of contamination on the property or belief that the property is contaminated. Stigma simply makes the property less desirable to a prospective purchaser, tenant, or visitor. Stigma is independent of the actual remediation costs to clean up the property and remains even after the property is cleaned up to meet governmental requirements. Stigmatized asset valuation is a common strategy used to develop compensatory awards in court cases involving environmental contamination. For example, in a California case, *Bixby Rance Co. v. Spectrol Electronics Inc.,* a jury awarded stigma damages to the owner of contaminated property even after the tenant at the property who was responsible for the contamination had paid for the cleanup.[22]

In a Minnesota case, damages were awarded for nuisance to proximity to landfill.[23] Findings in this case established that demonstration of physical damage to the property is a precondition to the award of stigma damages. In addition to damage awards to cover the costs to repair and restore the environment to its pre-contaminated setting, the court awarded damages to compensate for the loss in value of the property due to the stigma of contamination.

---

[20] *United States of America vs. Exxon Corporation, et al.,* United States District Court, District of Alaska, civil action no. A91-082 CIV, http://www.evostc.state.ak.us/History/Downloadables/GovMemo. pdf (accessed October 25, 2007).

[21] The oil company has other appeals pending related to questions of whether or not punitive damages are warranted and whether or not the $5 billion award is excessive. See Jim Clarke, "Judge Rejects Exxon's Request for a New Trial in Spill Trial," *Associated Press,* August 2 1998, http://www. cnn.com/2000/LAW/scotus/10/02/scotus.exxonvaldez.01/ (accessed October 25, 2007).

[22] *Bixby Rance Co. v. Spectrol Electronics Inc.,* California Superior Court, Los Angeles County, no. BC052566, December 1993.

[23] *Frank v. Environmental Sanitation Management Inc.,* 687 S.W. 2d 876, 883 (Mo. 1985).

In a recent Indiana court determination (the June 2001 decision in *Terra Products, Inc. v. Kraft General Foods, Inc.*) involving PCB contamination, the court ruled that where a plaintiff can demonstrate that repairs to real property do not restore the land to its former value, the plaintiff may recover both the costs of restoration and damages for any reduction of value that remains after remediation. In this case, the cost to remediate the property far exceeded the value of the property. The court also noted that legal requirements to remediate contaminated land apply without regard to cost. A plaintiff in Indiana now may be able to recover damages for the stigma attached to property following remediation if it can establish that the cleanup did not restore the property to its precontamination value.

Stigma damages have also recently been valued in relation to the storage of nuclear wastes. When the U.S. Congress passed the Nuclear Waste Policy Act of 1982 and amendments to the act of December 1987, it recognized the potentially significant socioeconomic dimensions of siting, constructing, and operating facilities for the storage and disposal of high-level radioactive wastes. Specific provisions were written into the act to enable prospective host states, tribes, and local governments to carefully and comprehensively assess socioeconomic impacts associated with waste disposal activities. Studies associated with the impact analysis assessment for locating a high-level radioactive waste facility at Yucca Mountain in Nevada found significant negative effects relating to loss of income from an actual or potential tourist industry.

In one recent study by the Nevada Agency for Nuclear Projects, the repository was compared in economic terms with a representative economic initiative that could be at risk if negative impacts do occur. A model of projected revenues associated with the Yucca Mountain repository was compared to a representative sample of the state's tourist industry, namely the visitor/gaming industry. Research addressed the questions: What would be the effects on the state economy if, as a result of the repository, one large hotel/casino project were canceled or chose not to locate in Las Vegas? What are the costs of losing such a project compared to with the jobs and revenue associated with the proposed Yucca Mountain repository? The analysis showed that, should the repository cause just one hotel/casino project not to locate in Nevada in the future, the immediate impacts to southern Nevada could be upward of 14,200 jobs and almost $500 million in revenue lost to the local economy annually.

Human Subject Experimentation

The above case summaries are a few of the precedents that support the people of Rongelap and their claims for hardship and related damages relating to involuntary resettlement, the loss of land and other critical resources, and the long-term impacts of environmental contamination of natural resources. With reference to the hardships, indignities, pain, and suffering associated with human subject experimentation, a number of settlements in U.S. cases are relevant.

In a negotiated settlement announced in November 1996, the U.S. government agreed to pay $4.8 million to twelve plutonium experiment subjects or their descendants ($400,000 apiece). In a medical research experiment at the University of Rochester, these subjects had received plutonium injections and were deliberately not informed about their involvement in the experiment and the contents of the injections.[24]

In December 1997, the Massachusetts Institute of Technology and Quaker Oats announced a $1.85 million settlement in a class-action suit representing the claims of more than one hundred former students. In the early 1950s, while living at the Fernald School in Waltham, Massachusetts, the students (ages twelve to seventeen) were given radioactive iron in their breakfast cereal once a month for three months.[25]

On May 27, 1998, U.S. district court judge John Nixon gave preliminary approval to a $10 million settlement in a class-action lawsuit against Vanderbilt University brought by 829 women who unknowingly had ingested radioactive iron while pregnant.[26] In 1946 the university had conducted a nutrition study involved 1,600 pregnant women, with 829 of them receiving a single dose of Fe-59 during a second prenatal visit before receiving routine therapeutic iron. On the third prenatal visit, blood was drawn and tested to determine the percentage of iron absorbed by the mother. The infant's blood was also examined at birth to determine the percentage of radioiron absorbed by the fetus. The doses were estimated in a March 1951 *American Journal of Obstetrics and Gynecology* article by Dr. P. Hahn to be between 5 and 15 rad. Contemporary estimates of the fetal dose by ACHRE and others suggest that it was a few hundred millirems. While a 1963–1964 study by Vanderbilt researchers found no significant differences in malignancy rates between mothers in control and exposed groups, it reported a higher number of malignancies among exposed offspring, with four children dying of childhood cancers. Researchers concluded a causal relationship between prenatal exposure to Fe-59 and cancer.[27]

---

[24] Advisory Committee on Human Radiation Experimentation, *Final Report*, 243–46. Award settlement reported in Melissa B. Robinson, "U.S. Reaches $4.8 Million Settlement with Radiation Victims," *Associated Press*, November 19, 1996, http://www.cipi.com/settlement.shtml (accessed April 1, 2008).

[25] Experiment summarized in Advisory Committee on Human Radiation Experimentation, *Final Report*, 342–44. Award settlement reported in "MIT, Quaker Oats to Settle Radiation Experiment Suit," *CNN*, December 31, 1997, http://edition.cnn.com/US/9712/31/radioactive.oatmeal/ (accessed April 1, 2008).

[26] "National News Briefs; $10 Million Settlement in Radiation Suit," *New York Times*, May 29, 1998, http://query.nytimes.com/gst/fullpage.html?res=9801EEDD1138F93AA15756C0A96 E958260&n=Top/Reference/Times%20Topics/Subjects/T/Tests%20and%20Testing (accessed December 18, 2007).

[27] ACHRE, *Final Report*, 348–49.

On May 5, 1999, a settlement was announced in the Cincinnati radiation case. Defendants paid $5.4 million but admitted no wrongdoing. In this case, cancer patients (mostly African Americans who were categorized as below-average intelligence and were charity patients) were exposed to large doses of whole-body radiation as part of an experiment sponsored by the U.S. military between 1960 and 1972. None of the subjects gave informed consent; they were told they were receiving treatment for their cancer. Subjects experienced nausea and vomiting from acute radiation sickness and pain from burns on their bodies. Some died prematurely as a result of radiation exposure.[28]

In the case of human radiation experiments on Alaskan Indians and Eskimos (1955–1957), involving administration of one single oral dose of iodine-131 to 102 native men, women, and children, public acknowledgement by ACHRE that a wrong had been committed eventually produced a formal apology from U.S. Air Force secretary F. Whitten Peters. Peters personally signed apology letters that were presented to Alaskan subjects and their families in October 2000. An announcement of a damage award settlement of $7 million accompanied public apologies (awards were made through the U.S. Department of Defense appropriation for 2001). The award includes payments of $67,000 to each study participant or his or her descendants as compensation for perceptions of health risk associated with an oral dose of iodine-131, especially the anxiety and stress caused by learning of the radioactive dosage. The North Slope Borough will also receive some $1.36 million as an award to the community.[29]

Establishing the "Value of Life"

Awards for damages in these cases generally acknowledge the difficulties of identifying an appropriate monetary value for the pain, suffering, and hardships of those whose lives were fundamentally altered by medical research abuses. Recently, the U.S. Environmental Protection Agency (EPA) established the value of statistical life (VSL)—that is, the monetized benefit from regulations that protect citizens from environmental risks known to produce cancers with high mortality rates. In studies supporting newly established regulations for levels of arsenic in drinking water, EPA conducted wage-risk studies using national income data. Assuming a 26 percent mortality rate for bladder cancer and an 88 percent mortality rate for lung cancer, the current VSL used by the EPA is $6.1 million per person, in 1999 dollars. This value does not reflect any adjustments to account for national real

---

[28] Cincinnati Radiation Litigation, 874 F. Supp. 796 (S.D. Ohio 1995).
[29] The study involved two hundred administrations of I-131 to 120 subjects, including 19 Caucasians, 84 Eskimos, and 17 Indians (Advisory Committee on Human Radiation Experimentation, *Final Report*, 598–603). Award settlement reported in Sam Bishop, "Feds, Natives Settle over '50s Iodine Experiments," *Fairbanks Daily News-Miner,* October 26, 2000.

income growth occurring subsequent to the completion of the studies. Were the agency to adjust the VSL to account for this growth, it would be approximately $6.77 million (assuming a 1.0 percent income elasticity).

EPA arsenic regulations also establish a willingness-to-pay value for those cancer cases that do not result in mortality. The willingness-to-pay value for avoiding a nonfatal condition is based on economic research that established a willingness-to-pay value to avoid a case of chronic bronchitis. EPA noted that use of this proxy might understate the true benefit if the willingness to pay to avoid a nonfatal cancer is greater than the willingness to pay to avoid a case of chronic bronchitis. The mean value of this willingness-to-pay estimate is $607,000 per person (in 1999 dollars).[30]

Recommendations for Categories of Concern in This Claim

Some might argue that the social and cultural changes, physical injuries, and economic hardships experienced by the people of Rongelap during evacuation periods and after leaving their atolls in 1985 reflect the broader experience of modernization and development in the Pacific. Modernization and development have swept through the entire Pacific region. Accompanying these changes has been erosion of traditional customs and relationships in the face of a market economy and a consumer lifestyle, dietary shifts, and increased health problems associated with a Western diet. Since World War II, the Pacific region has seen demographic shifts, with the population of many island nations living in densely populated urban settings. Environmental degradation, dietary changes, and resulting degenerative health conditions (including diabetes and hypertension) have accompanied these changes.

However, as a direct result of the U.S. nuclear weapons testing program, the pace of change and the nature of change are radically different for the Rongelapese as compared to those of the broader Pacific region. The events that resulted in involuntary evacuation in 1946 and 1954 were sudden, systemic assaults that forcibly evicted the community from a traditional way of life and thrust it into a life forever bound to Western culture, food, medicine, and technology. The 1982 disclosure that subsistence lifestyles in Rongelap posed serious threats to the health and safety of the people of Rongelap produced an immediate fear and led to the 1985 exile of the community, against U.S. government wishes and without U.S. government assistance. Loss of a healthy environment and loss of all access

---

[30] United States Environmental Protection Agency National Primary Drinking Water Regulations; Arsenic and Clarifications to Compliance and New Source Contaminants Monitoring; Final Rule [40 CFR Parts 9, 141, and 142]. Federal Register: January 22, 2001 (Volume 66, Number 14: 6975–7066).

to Rongelap, Rongerik, and Ailinginae, coupled with the difficulties of life in fragmented communities on leased land, further increased community reliance on Western food, culture, and technology and produced a redefined notion of what was good and bad. The Rongelapese adjusted to these events and conditions without the luxury of time, community integrity, and the related resources that other Pacific island nations enjoyed in the last half of the twentieth century. Today, the Rongelapese live in fragmented communities characterized by anxiety and melancholy for what has been lost and what, for future generations, may never again be enjoyed. Other Pacific Islanders share some of their conditions and experiences, but the Rongelapese experience is one of extremes.

Reparations—meaningful efforts to repair or make amends for the individual assaults and cumulative injuries experienced by individuals, the Rongelap community, and the nation over many decades—could be addressed through various actions designed to provide restitution, indemnity, and satisfaction under four general categories of concern. These categories are:

- The hardships, injuries, and consequential damages of the loss of a healthy, self-sufficient way of life
- Natural resource damage and related socioeconomic stigmatization
- Consequential damages of human exposure to fallout from the U.S. nuclear weapons testing program
- Negligence, negligent misrepresentation, battery, and related consequential damages of involuntary participation in human subject research.

Category of Concern 1: Loss of a Healthy, Self-Sufficient Way of Life

One of the major consequences of the U.S. nuclear weapons testing program was the eventual loss of a healthy, self-sufficient way of life previously enjoyed by the people of Rongelap, Rongerik, and Ailinginae. The loss of the material basis to sustain a healthy, self-sufficient way of life produced social, cultural, economic, and political hardships and injuries. Damages include, and specific remedial actions are needed to address:

- Social, cultural, economic, and political hardships, indignities, and other injuries resulting from evacuation of Rongelap and Ailinginae and involuntary resettlement in 1946
- Social, cultural, economic, and political hardships, indignities, and other injuries resulting from evacuation of Rongelap and Ailinginae and involuntary resettlement from 1954 to 1957.
- Social, cultural, economic, and political hardships, indignities, and other injuries resulting from evacuation of Rongelap and Ailinginae in 1985 and involuntary

resettlement in Majuro, Ebeye, Mejatto, Hawaii, and other communities beginning in 1985 and continuing through an unknown future date.
- Social, cultural, economic, and political injuries resulting from loss of access
  to and use of, and the extensive contamination of, sacred land (Rongerik Atoll)
  from 1946 through an unknown future date when the environment (lagoon,
  reefs, islands, and all resources within contaminated by radioactive and other
  toxic materials used in U.S. nuclear weapons tests) has been restored.
- Social, cultural, and economic injuries resulting from the loss of access to and
  safe use of terrestrial resources in Rongelap and Ailinginae from 1954 through
  an unknown future date when the environment—soil, water, vegetation, and
  other life on all atoll islands contaminated by radioactive and other toxic materials
  used in the U.S. nuclear weapons testing program—has been restored.
- Social, cultural, and economic injuries resulting from the loss of access to and
  safe use of marine resources in Rongelap and Ailinginae from 1954 through
  an unknown future date when the marine environment (lagoon, reefs, and all
  resources within contaminated by radioactive and other toxic materials used in
  U.S. nuclear weapons tests) has been restored.

In the Marshall Islands, land is not a commodity that can be bought or sold; it is
an integral component in the "natural capital" that sustains a way of life.[31] In Western
property law, loss of land means loss of income. But for the Marshallese, and for
many traditional cultures where power, authority, status, and meaning is shaped
by kinship and the meaningful use of communally owned resources, loss of land
represents the loss of the means to sustain social institutions, reinforce kinships
systems, and survive and thrive as a self-sufficient entity. While the Marshallese
generally refer to inherited rights as land rights, land includes the lagoon, reefs, clam
beds, and broader atoll ecosystem, as well as the marine, arboreal, and terrestrial
life within. Thus assessment of damage and compensatory actions need to reflect
the damages resulting from the contamination and loss of the "natural capital" that
sustained a way of life.

The RMI Nuclear Claims Tribunal has established an approach for compensating
for loss of land based on dry land acreage estimates and "market" assessments of lease
value. This valuation represents one portion of the economic value of damaged and
destroyed atoll ecosystems. In the cases of Enewetak and Bikini, the tribunal ruled
that land-value awards included compensation for the loss of access to and use of
natural resources contained within (palm trees, pandanus, and so forth). Missing
from the equation were marine resources, cultural resources, and the sociocultural
consequences resulting from the loss of land.

---

[31] As discussed in Lansing, Lansing, and Ezro, "Value of a River."

To adequately compensate for damage and loss of marine resources, acreage estimates for loss of land could be expanded to include lagoon and surrounding reef acreage, with damage awards paid to the entire community in the form of a trust. Income from the trust could be used to develop the internal capacity to study, improve, reduce environmental risks to, and eventually restore degraded ecosystems. Subsistence use value has been partially determined via the formula used to calculate lease value in the loss of land. Applying the same formula to include lagoon, reef fringe, and broader marine acreage surrounding the seamount just north of Rongelap would produce a value representing loss of marine resources. Calculations for lagoon areas are:

- Ailinginae lagoon area: 105.96 square kilometers, or 26,183.14 acres
- Rongerik lagoon area: 143.95 square kilometers or 35,570.62 acres
- Rongelap lagoon area: 1,004.32 square kilometers or 248,171.49 acres[32]

Loss of access to Rongerik and subsequent involuntary relocation from Rongelap and Ailinginae generated significant political consequences: representation of the three atolls was reduced to just one senator, and because the community was so dispersed, the atolls did not receive the same level of economic development in subsequent years. Benefits of representation include increased services provided by the RMI government. To adequately compensate for this and other damages, including the loss of cultural resources and the sociocultural consequences resulting from loss of a healthy way of life, remedial awards should reflect a commitment to restore the community to a self-sufficient and meaningful way of life, with damage awards paid to the community in the form of a trust. Income from the trust could be used to address the sociocultural, job training, economic development, and related needs of the community identified in part 4 of this report. Such an award could be based on the combined estimated cost of capital infrastructure common to other atolls (electricity, telephones, schools, medical clinics, transportation facilities, public water systems).

Category of Concern 2: Natural Resource Damage and Related Stigmatization

Natural resource damage from the U.S. nuclear weapons testing program is severe, toxic, and persistent, representing permanent damage that reduces the value and potential future use of Rongelap, Rongerik, and Ailinginae. Contamination from

---

[32] Lagoon area figures are taken from maps posted on the RMI Embassy website, http://www.rmiembassyus.org/Indiv%20Atolls/Republic%20of%20the%20Marshall%20Islands%20-%20Rongelap%20Atoll.htm.

nuclear weapons testing fallout—including tracer chemicals such as arsenic, cadmium, and other mineral isotopes used to fingerprint the fallout from each weapon—has damaged natural resources and generated socioeconomic stigmatization. Degenerative and stigmatized conditions that affect current and future resource values and economic development potentials have not been addressed in awards for loss of property and cost to restore.

It is recommended that compensatory actions incorporate principles of nuclear stewardship and provide sufficient funds, facilities, expertise, and training to give the people of the Marshall Islands the means and ability to conduct their own intergenerational epidemiological surveys and environmental risk assessments and to develop culturally appropriate environmental risk-management strategies (including monitoring contamination levels and decay rates and remediating terrestrial and marine ecosystems). Shaping an effective program of nuclear stewardship implies local control of funds and a level of effort that reflects the best practices and strategies currently used to assess, remediate, and/or manage waste sites in the United States. Fiscal responsibility to initiate and implement a meaningful nuclear stewardship program should lie in the hands of those who transformed the Marshall Islands from a healthy, viable atoll ecosystem into a nation pockmarked with nuclear waste sites.

A nuclear stewardship program should assess the broad range of environmental hazards persistent in the terrestrial and marine ecosystem, address natural resource damages, and provide some measure of remedy for socioeconomic stigmatization.

Natural resource damage includes injuries resulting from involuntary resettlement and the inability to defend commercially significant resources from foreign exploitation, as well as stigma damages associated with the radioactive contamination of the region. Stigma damages relevant to the nuclear weapons contamination of Rongelap, Rongerik, and Ailinginae atolls include perceptions of hazard with reference to locally produced foods and handicrafts and locally harvested marine resources, and perceptions of hazard discouraging ecotourism industry growth.

One strategy to determine the value of natural resources is to consider market, nonmarket, and subsistence values for natural resources in similar settings. A 1997 study, *The Economic Value of Majuro Atoll's Coastal Resources,* prepared for Majuro Atoll's local government and the Coastal Management Program by economist Philip King, provides a proxy example. King identified a range of market values for coastal-zone resources, including pelagic and reef fisheries, tourism, handicrafts (shells and pearls), copra, agriculture, and sand and gravel mining. The total was $161,976,377.25. The resources of three atolls would be worth considerably more, as would the value of localized resources (giant clam beds, valuable golden cowry shells, and the major source of arrowroot). Values would need to be adjusted to reflect lost opportunity over the past fifty years, as well as lost future opportunities (for example, for thirty years, at which time environmental contamination could be reassessed and projected losses recalculated).

A relevant case example of assigning a value to nonmarket uses of a beach is contained in the September 1999 final settlement in a Huntington Beach, California, oil spill. In 1997 a jury verdict set legal precedent by awarding a total of close to $13 million in damages for the loss of public recreation. The award set a $13.19 per-capita value of the lost enjoyment for one day at the beach.[33] Applying this example to the Rongelap claim, and using the California figure as a proxy for recreational value, damages for the entire Rongelap population for past, current, and future loss of recreational value can be calculated. If, for example one hundred people lost a total of 1,760,000 days over the past fifty years at $13.19 per day, the total value of lost recreational access comes to $23,214,400. If four thousand people will lose 4,224,000 days over the next thirty years at $13.19 per day, the total projected value of lost recreational access comes to $557,145,600.

A "subsistence" value of natural resources was suggested in the 1994 Marshall Islands report *Strengthening of Agricultural Support Services,* written by Ray Shaw for the Asian Development Bank. His figures suggest that the average value of household production equals approximately $2,585 (1997 values).[34] This subsistence value represents the value of household production to put food on the table. This value does not incorporate the value of other subsistence activities that provide housing, medicine, and household, artisanal, and craft supplies, including baskets, mats, shell necklaces, toys, nets, boats, and tackle.

In calculating appropriate awards for stigma damages, it is important to identify the monetary value of the highest and best use of the land. Possible categories include agrarian and residential use, tourist use, and—given the level of contamination—temporary storage of nuclear waste.

For survival without overusing the resource base, sustainable subsistence production requires access to multiple locations. Cleanup and resettlement of the main island of Rongelap Atoll is not sustainable without restoration of all Rongelap's

---

[33] Attransco (owner of the tanker) appealed, lost, and agreed to settle for $16 million in damages paid to the state of California. Attransco will also pay the California State Department of Fish and Wildlife $5.3 million for harming plankton and other "tiny sea creatures" that live along the coastline. In addition, BP America and the Trans-Alaska Pipeline Liability Fund previously paid $7 million to settle their role in the case. The owner of the dock, Golden West Refining Company, paid $4.5 million to settle. BP America also spent $12 million to clean up the oil spill.

[34] Shaw examined primary production by households for consumption between 1994 and 1997. Categories considered included agricultural products (bananas, breadfruit, pandanus, green and mature coconut, taro, other vegetables, fruits); meat products (pork and chicken); and fish and shellfish (including crabs). Estimated value (using 1991 and 1992 national income estimates and the 1988 census of population and housing): agricultural products: $1,372,425; meat products: $2,050,540; fish and shellfish: $3,747,476. These figures assume a national total of 2,401 households growing food; 3,399 households engaged in fishing; and 2,253 breeding livestock.

islands, as well as Rongerik and Ailinginae atolls. However, current scientific knowledge limits the ability to safely decontaminate and restore natural environs. Current and future losses associated with "off-limit" islands might be extrapolated from the sale of islands in other parts of the Pacific.

An alternative approach would be to consider atoll areas that cannot be restored to pretesting conditions with existing technology, including lagoons, islands, and reefs, as de-facto nuclear waste storage facilities. Compensation for this de-facto status is warranted. Several measures of value could be used to determine the appropriate compensatory award. These measures include:

- Estimation of the total quantity of contaminated areas designated as too contaminated to restore
- Estimation of the quantity of low-level, mid-level, and high-level radioactive waste
- Rates charged at temporary storage facilities for these types of materials
- Damages to reflect the socioeconomic stigma associated with hosting radioactive storage facilities in the northern Marshall Islands.

Category of Concern 3: Consequential Damages of Human Exposure to Radiation

"Exposed" people of Rongelap include those living on Rongelap and Ailinginae in 1954 who were exposed to Bravo and contamination from earlier tests; those who were resettled in 1957; those who were born on the contaminated atoll; those who were exposed to materials and food originating from Rongelap, Rongerik and Ailinginae atolls; and the descendants of people exposed to radioactive contaminants. Given the synergistic, cumulative, and genetic effects of long-term exposure to radioactive isotopes and other environmental contamination from military testing, exposure is of concern to this and future generations. Psychosocial stigmatization, pain, indignities, and other suffering resulting from exposure to radiation and other contaminants from the U.S. nuclear weapons testing program were endured by the community in the past, continue to trouble them in the present, and, given the nature of these injuries, will assuredly present hardships in the future.

The U.S. nuclear weapons testing program effectively took the land and health of the Rongelapese and altered the future of the community of Rongelap (and other Marshallese communities) by creating a severely contaminated environment. This taking occurred without just compensation. In returning the Rongelapese to a contaminated setting and failing to notify them of the risks associated with residence in Rongelap, the U.S. government failed to perform its public duty to protect the health and way of life of its trustees. At the same time, the U.S. government and its citizens benefited from use of the Marshall Islands as a nuclear testing site, including the development of a national security system and the achievement of nuclear détente.

Exposure to environmental hazards generated by the U.S. nuclear weapons testing program resulted in stigmatization and other psychosocial injuries that have adversely affected individuals, the community, and the nation. Nuclear testing introduced new taboos: certain lands and certain foods were off–limits; marriage to certain people involved new social stigmas; birthing presented new fears and health risks; family life often involved the psychological, social, and economic burden of caring for the chronically ill and disabled. The failure of the U.S. government to provide the Rongelap people with adequate and accurate information concerning environmental hazards and risks, coupled with contradictory pronouncements on what was and was not safe, created taboos that were incomprehensible yet permeated all aspects of life. This imposition of new taboos effectively transformed the loci of control over taboos: undermining the rules and customary power structures that shape, interpret, and reproduce strategies for living in the Marshall Islands. Furthermore, the fear of nuclear contamination and its adverse effect on the health of current and future generations colors all aspects of social, cultural, economic, and psychological well-being. Nuclear taboos orchestrate people's behavior and use of their land and resources.

This social stigmatization adversely affects the economy, society, family, and individual health and well-being of the people of Rongelap, Ailinginae, and Rongerik, and to varying degrees the nation as well. Broadening the recognized categories of "exposed" and the recognized modes of exposure will increase the number of potential claimants for personal-injury damage and medical treatment. Awards for consequential damages of radiation exposure should reflect the experiences, damages, and treatment needs of individuals, households, the community, and the nation. Damages could be calculated with consideration of the benefits that the United States received from the testing program and subsequent creation of what in legal terms is referred to as a nuisance.

Category of Concern 4: Negligence, Negligent Misrepresentation, Battery, and the Related Consequential Damages of Involuntary Participation in Human Subject Research

The complex biophysical, sociocultural, and psychological assaults on the people of Rongelap outlined in this report are problems that no single community can surmount without intensive assistance that is first and foremost focused on the needs and concerns of those who have been harmed. This report not only documents U.S. culpability for damages and loss but also presents evidence in support of the argument that the U.S. government was negligent in providing adequate and meaningful remedies. This negligence includes sponsoring and implementing a biomedical research program involving the purposeful exposure of Rongelapese to radioactive and toxic hazards that further compromised already damaged immune systems and further inhibited individual and community abilities to adapt and adjust. This research was based on questions of scientific interest and military concern rather than a holistic approach to individual therapeutic needs of a severely exposed population.

The elements of negligent misrepresentation are: (1) a duty by the defendant to communicate accurate information to the other party; (2) a false statement of material fact or carelessness in ascertaining the falsity of the statement by the party making it; (3) intention to induce the other party to act; (4) action by the other party in reliance of the statement; and (5) damage to the other party from such reliance.

Negligent misrepresentation occurred with the repeated false assurances by the United States that:

- Nuclear weapons tests between 1954 and 1957 did not result in an increased contamination of the islands.
- The Rongelapese were being returned in 1957 to a safe environment.
- Nuclear weapons tests after 1957 represented no danger of additional fallout to the Rongelapese.
- A subsistence-oriented way of life on Rongelap (1957–1985) did not threaten the health of the population.
- Arrowroot, clams, coconut, pandanus, fish, and other foods were not contaminated and were safe to consume.
- Incidents of poisoning from exposure to contaminated foods were the normal results of inadequate food storage or preparation and not the result of contaminants.
- Reproductive health problems, including high rates of miscarriage and gross birth defects, were not related to radioactive exposure and were normal occurrences for a Marshallese population.
- Chronic incidents of degenerative disease appearing after 1954, including diabetes, cataracts, and various immune system problems (such as skin lesions that failed to heal), were not related to initial or cumulative exposure and thus were not the treatment responsibility of the medical survey (and later) 177 Agreement health programs.
- Radioactive exposures produced no genetic effects.
- Medical survey exams, procedures, and various experiments were necessary elements of a therapeutic treatment program and were in the best interest of individual patients.

Misrepresentations were made to ensure continuation of experiment conditions—a bounded unit of analysis living in a controlled, isolated, intensively studied setting—and continued access to valued research subjects. Misrepresentations resulted in Rongelapese compliance with the 1957 resettlement and their extended stay on Rongelap until the 1985 evacuation. Misrepresentations resulted in the continued involvement of Rongelap subjects in medical survey exams, procedures, harvesting of samples, and various human subject research experiments. These involvements contributed to an establishment of minimum-dose thresholds, furthering understanding of the way radionuclides move through the food chain and through the human body and furthering understanding of the various human effects of prolonged exposure to radiation. The U.S. military needed this information during the Cold War so that it could ascertain how well troops and citizens would survive in the event of nuclear war. Misrepresentations induced significant injury, indignities, and other harm on the people and community of Rongelap and produced findings of intense scientific interest and value to the U.S. military.

Returning an acutely exposed population to a contaminated setting for the purpose of documenting the long-term effects of previous and repeated exposures to radioactive substances constitutes battery or purposeful physical assaults on the people of Rongelap, Rongerik, and Ailinginae. Using this population in various nontherapeutic human subject experiments that involved exposures to radioisotopes, whole-body irradiation, and other procedures constitutes battery or purposeful physical assaults on the people of Rongelap, Rongerik, and Ailinginae.

Conducting research exams, failing to share findings, withholding medical records, harvesting samples, and subjecting the Rongelapese to invasive procedures as part of human subject experiments constituted abusive indignities that violated cultural norms and violated individual rights to informed consent. These assaults constitute physical battery.

Nuclear testing created environmental hazards and health problems that will persist for decades to come. Remedial actions are needed to address the consequential damages of U.S. government negligence in the form of a medical treatment program that adequately and effectively meets the acute, chronic, and degenerative health care needs of the people of Rongelap and other communities exposed to radiation and other toxic materials produced by nuclear weapons tests. Compensatory damages for abuses associated with experiments on human subjects should be awarded to provide some measure of remedy for the pain, suffering, and indignities endured by the Rongelap community as a result of their:

- Involvement in long-term human ecology studies on the effects of radiation beginning in 1954 and extending through 1997

- Individual, community, and national injuries resulting from willful misrepresentation of the level of contamination, extent of contamination, and safety of living, eating, and reproducing in a contaminated setting
- Use as human subjects in a range of isolated experiments, including chromium-51 and other radioisotope experiments that had little or nothing to do with individual health and treatment needs
- Inability to exercise their rights to meaningful, informed consent.

Concluding Remarks

In its "Memorandum of Decision and Order" in the class-action claim for and on behalf of the people of Enewetak, and in its "Memorandum of Decision and Order" in the class-action claim for and on behalf of the people of Bikini, the RMI Nuclear Claims Tribunal has established parameters for compensating the hardships and related consequential damages of the U.S. nuclear weapons testing program. In the cases of Enewetak and Bikini, consequential damages were awarded for hardships consisting of severe food shortages and hunger, disease, loss of culture, and other types of pain and discomfort, with damages reflecting the cumulative total of individual awards that do not exceed existing bodily injury awards granted by the tribunal.

This report has presented evidence to support claims of hardship and consequential damage experienced by individuals *and* the corporate community of Rongelap. Evidence has been presented to demonstrate a wide range of abuses in addition to the hardships of loss of land and involuntary resettlement (severe food shortages and hunger, disease, loss of culture, and other types of pain and discomfort). These additional damages include natural resource damages, socioeconomic stigmatization, intergenerational and psychosocial damages of exposure, and the physical and psychological damages accompanying human subject experimentation.

This report also presents evidence of similar instances of damage and abuse that have produced compensatory awards in the United States that far exceed the existing ceiling set by the tribunal. In many of these cases, especially those pertaining to human subject experimentation, awards represent remedy for single acts of abuse, as opposed to this case of multiple and cumulative abuses experienced by individuals and the community over the span of many decades. It is our contention that compensatory actions are warranted for each of the individual incidents and abuses outlined under four general categories of concern, and that these damages should reflect the individual as well as corporate experience of:

- The hardships, injuries, and consequential damages of the loss of a healthy, self-sufficient way of life

- Natural resource damages and related socioeconomic stigmatization
- Consequential damages of human exposure to radiation and fallout from the U.S. nuclear weapons testing program
- Negligence, negligent misrepresentation, battery, and related consequential damages of involuntary participation in human subject research.

Epilogue

Seeking Meaningful Remedy

In late October 2001, the Rongelap community assembled on Majuro, the capital of the Marshall Islands, to attend the Nuclear Claims Tribunal hearing and to collect their personal-injury awards, which are paid out each year in annual pro rata payments. They traveled from their homes on Mejatto, a small island in Kwajalein Atoll; Ebeye, across from the main island of Kwajalein; outer islands; and Hawaii. They stayed with Rongelap family and friends who had settled on Majuro.

We arrived in Majuro a few days before the October 29–November 2 hearing to meet with the community and prepare our presentation of the case. At the Rongelap local government offices, we found the corridors and rooms filled with people who were catching up, getting help for various problems, working on baskets and jewelry, and cooking food for community meals. Outside the office, the women of Rongelap met each evening to sing and dance, practicing song and movement that told their story of struggle and survival. *Alab* John Anjain had brought his community record books from his home on Ebeye. Using the Rongelap local government copy machine, we reproduced his maps depicting land claims and land-use history to use as exhibits.

We prepared other exhibits: To demonstrate that "land" from a Marshallese perspective includes lagoons and surrounding reefs, we made copies of a series of maps for each atoll that had previously been marked up to illustrate the names and locations of sacred sites and critical resources such as freshwater wells, giant clam beds, and significant reefs. To put a human face to the history and the numbers, we asked the Rongelap people to help us compile lists of names, identifying the original eighty-two people (four of whom were pregnant) severely exposed and evacuated in 1954; the twenty people in that group who have since died from radiation-related illnesses; the eight women who have gone on record as suffering miscarriages or giving birth to children with severe congenital defects (such matters are generally taboo, and it is a difficult topic to discuss publicly); the thirty-five children born with severe congenital defects (and the names of their parents); and the nineteen people who suffered from a preventable epidemic of polio. We also made copies of photographs: images of children, the elderly, life on Rongelap, and examples of congenital defects. We posted these images on the NCT hearing room walls to remind us of those people who had suffered greatly and who could not be there to tell their stories. We met with individual members of the Rongelap community, talking with people for hours on end about their knowledge and experiences, answering their questions, and identifying which community members might best be able to articulate their understandings and expertise and thus demonstrate the elements of our expert witness report.

The tribunal hearing on the Rongelap claim took place over a three-day period, beginning on Halloween and extending through All Saints' Day and All Souls' Day. Halloween and the Christian holy days of obligation, All Saints' Day and All Souls' Day, have relatively little social meaning in the Marshalls (a few kids go trick-or-treating on Majuro, but this is a relatively recent activity). But for those raised in a Catholic tradition, it seemed entirely appropriate to consider the horrific consequences of nuclear war games in the Marshalls on days such as these—days devoted to honoring and remembering the dead.

Before the hearing, a reverend member of the Rongelap community offered prayers and blessed the gathering. The proceeding was formally opened with remarks by the NCT judges. Iroij Mike Kabua welcomed the Rongelap community. Senator Abacca Anjain-Maddison read a statement written by John Anjain (former magistrate). Councilwoman Rokko Laninbelik read the names of those in the Rongelap community who had survived the 1954 exposure and those who had died. The expert witnesses for the Public Advocate and the Defender of the Fund were introduced, expertise was established, we were sworn in, and our reports were formally received into the evidentiary record. We were then invited to present our case.

The room was hot, with bright florescent lights and hard plastic seats. It was a big room full of people, and every chair seemed to be occupied. This modern realm seemed miles away from the Pacific reality outside our doors: ocean, sand, palm trees, a balmy breeze over reef and lagoon. But from the moment we began to present the case, all ears tuned in, time shifted, and some seventy-five or so people—anthropologists, lawyers and advocates, judges, clerks, translators, technicians, the press, and the Rongelap community—were immersed in an intense and at times intensely surreal experience.

We began our case with a request: Johnston would present several summaries of the core issues and events, and Barker would ask questions of fourteen Marshallese witnesses to illustrate and further contextualize our summations. We asked the tribunal to accept Marshallese testimony as evidence rather than anecdotal material, arguing that such testimony was substantiated by the documentary record cited in our report *and* that it represented direct evidence of the oral history and customary traditions of a nonliterate society. To support this request, we cited the Zuni land-claims case, a U.S. federal court precedent that allowed oral history testimony to be entered as material evidence in a land claim. We also asked the tribunal to consider Marshallese customs, especially the need for questions about taboo subjects to be phrased in the Marshallese language, with nuanced respect, and the need for questions to women to be presented by women. Thus we asked the tribunal to allow us, the anthropologists, to directly question Rongelap witnesses who, we argued, were cultural experts. Other Marshallese witnesses testifying about current conditions or recollections of a nonintrusive or nontraumatic nature were to be interviewed directly by Public Advocate Bill Graham and Rongelap local government counsel John Masek. All witnesses appearing in support of the Rongelap claim would be available for cross-examination by Defender of the Fund Phil Okney. The tribunal expressed concern that the questioning of Marshallese witnesses by anthropological experts might violate procedural rules but allowed us to proceed, acknowledging the opportunity for us to demonstrate our ethnographic methods as well as the need to proceed with proper respect for customs and taboos. Counsel was asked to provide the tribunal with a post-hearing brief further outlining the rationale and legal precedents for this approach.[1]

The proceedings took place in English. With the help of translators, all questioning and testimony from Marshallese witnesses occurred in Marshallese with English translation. The hearings were tape-recorded and videotaped to ensure as complete a record as possible. Senator Abacca Anjain-Maddison also videotaped the proceedings.

Our summations and the testimony from Marshallese witnesses allowed us to establish a sense of what was—a culturally vibrant, traditional way of life based on access to and stewardship of critical terrestrial and marine resources—before moving on to the broader questions concerning what had happened, to what effect, and with what consequences. We heard testimony on land rights and stewardship and the many ways that regular access to and use of critical resources allowed elders to teach and transmit a way of life to the next generation. We heard testimony about the hardship of being forced to leave Rongelap during weapons tests. Testimony was given about the historical importance of Rongerik, its role in sustaining the

---

[1] In arguments to the NTC, Public Advocate Bill Graham, lead counsel for the Rongelap claim, noted that many of the witnesses were elderly women whose experiences had been very humiliating and painful, and that the public discussion of these experiences—in front of family members and a broader public—broke numerous taboos. Counsel asked the tribunal to recognize the cultural and linguistic necessity of anthropologists posing questions in ways that minimized cultural sensitivities, respected taboos, and elicited testimony (review of the Rongelap hearing videotapes for October 31, 2001). The post-hearing brief cites U.S. federal rules for evidence, State of California rules for questioning expert witnesses, case-study precedents involving land-claims cases (*Zuni Indian Tribe v. United States*), the Canadian Supreme Court ruling on *Delgamuukw v. Province of British Columbia*, and Australian court evidentiary procedures in cases involving native title, the role of anthropologists in these proceedings, and the standing of native elders as cultural experts. See the claimants' brief in support of their use of an anthropologist to question witnesses, filed with the NCT clerk, February 12, 2002.

Rongelap community, the damages that resulted when the navy (the Trust Territory administrator at the time) decided, without the permission of Rongerik landowners, to resettle the Bikini community there, and the efforts that the people of Rongelap took to help the Bikinians as resources dwindled and they faced starvation. And we heard poignant testimony about the psychosocial and economic stigmatization resulting from radiation exposure, the personal hardships and injuries associated with radiation exposure, and the personal hardships and injuries associated with serving as human subjects in medical experiments and research programs that lasted for decades.

A great deal of new information came out in memories shared by our Marshallese experts. Four areas are worth elaborating upon here, as this information was not addressed in any exhaustive form in *The Rongelap Report*.

First, the long-term medical research involved some seventy-three excursions to the Pacific to examine, document, and collect samples for human radiation research. This research involved the Project 4.1 people, who were on Rongelap during the Bravo test and were considered "exposed," and the people from Rongelap who were not there in 1954 but moved back after 1957 (the "nonexposed" Rongelapese used as research controls). The human subject research also involved hundreds of people from Utrik, Likiep, Majuro, and other atolls in the Marshall Islands who were selected as age-matched and "nonexposed" controls for various biomedical studies. As was the case for the people of Rongelap, control exams were conducted with little or no information given to the subjects, and until the mid-1980s, no effort was made to obtain informed consent. The traumatic nature of this experience is emphasized in the following excerpt from the testimony of Helena Alik, who was enrolled in the study as a control subject at age eleven.[2]

Holly Barker: *When you participated in these medical programs, did you understand the medical procedures or treatment that you were receiving?*

Helena Alik: *I have no idea. I only know that when the time comes for the people of Rongelap on Ejit to be examined, I am sure that I will also be called on to attend the examination.*

Holly Barker: *Have you ever had any fears or concerns about participating in this program?*

Helena Alik: *Sometimes I was afraid and hide, I would run away because I heard that, they say that I am going to attend the medical examination of those contaminated people.*

---

[2] Helena Alik, testimony recorded on videotape, Nuclear Claims Tribunal hearing on the Rongelap claim, November 1, 2001.

Holly Barker: *What would you want to do? Would they come and find you?*

Helena Alik: *I was always get caught. Sometimes I am in school in the classrooms and they send police officers.*

Holly Barker: *Did you ever feel embarrassed when the police came to the school, confused about why they were there?*

Helena Alik: *Yes, because there were children asking: "Are you also a victim of the fallout?"*

This and other testimony illustrated several facts about the medical research program that had not been previously considered within the tribunal or in the earlier ACHRE review. The Rongelap study was structured in ways that required the involvement of children from other atolls, especially children in the southern part of the nation. Such involvement extended over decades. Control subjects were selected at the direction of authorities. Being singled out resulted in social stigmatization (people were shunned because of the social perception that all people studied by the medical survey team were damaged by radiation). Control subject experiences included thorough examinations with photographs and x-rays; measurement of internal radiation with whole-body counters; the sampling of blood, bone marrow, skin, and other tissue; and, on a number of occasions, the injection of radioisotopes, vaccines, and other nonexplained substances. While the United States had formally acknowledged that chromium-51 and tritiated water had been given to Marshallese subjects as part of a nontreatment-related research program, medical technician testimony during the tribunal hearing noted that these and other radioisotopes were administered at various times to small groups of people prior to whole-body counts.[3] Radioiodine, for example, was administered as part of iodine-uptake studies. Testimony supported ACHRE findings that some

---

[3] As many as seventy-nine Marshallese subjects received Cr-51 and/or tritium as part of nontreatment-related research activities between 1961 and 1966, and between five and twenty-five subjects may have received Cr-51 during the 1959 medical survey. Several of these studies also involved Caucasian Americans who were living in the Marshall Islands. See "Total-Body Water and Hematologic Studies in the Pacific Islanders Using Chromium-51 and Tritium," *Department of Energy*, http://www.hss.energy.gov/healthsafety/ohre/roadmap/experiments/0491docb.html#0491_Brookhaven (accessed May 10, 2007); Leo M. Meyer, letter to W. Siri, April 9, 1965, Subject: Sending you the 1.0 volumetric pipette which we used at Kawjalein, http://worf.eh.doe.gov/data/ihp2/2522_.pdf (accessed April 11, 2007); Robert A. Conard, Leo M. Meyer, Wataru W. Sutow, William C. Moloney, Austin Lowrey, A. Hicking, and Ezra Riklon, *Medical Survey of Rongelap People Eight Years After Exposure to Fallout* (Upton, NY: Brookhaven National Laboratory, 1963), http://worf.eh.doe.gov/data/ihp1a/2682_.pdf; Robert A Conard et al., *Medical Survey of the People of Rongelap and Utirik Islands Eleven and Twelve Years after Exposure to Fallout Radiation (March 1965 and March 1966)* (Upton, NY: Brookhaven National Laboratory, 1966), http://worf.eh.doe.gov/data/ihp1a/2683_.pdf.

biomedical procedures had no relationship to the medical needs of individuals. And testimony made clear how the experience of serving as a research control was intrusive, painful, and potentially harmful to the health of the participant.

The second major area of new evidence examined at this hearing involved the fact that damages from U.S. nuclear militarism were borne by the entire nation, not just individuals, atoll populations, or specific landowners. And the damages associated with U.S. military testing include much more than damage from radioactive elements. These points—nationwide exposure and the broader array of damages associated with Cold War militarism—were most effectively made by Dr. Nancy Pollock, the anthropologist who testified on behalf of the Defender of the Fund, in her comments to the tribunal following our presentation of damages associated with a no-re-evacuation policy. To illustrate some of the damages associated with leaving the people of Rongelap on their atoll and exposing them to future tests—for the purpose of using them as long-term research subjects, as well as demonstrating to the world that they suffered no lasting harm from their exposure—we presented the tribunal with details from a 1968 series of biochemical weapons tests known as DTC 68-50, or Project Shad. This test series took place in September and October 1968 in the waters off Enewetak and involved the atmospheric dissemination of "PG"—staphylococcal enterotoxin B—an anthrax-simulating agent that produces immediate and extreme flulike symptoms that can be fatal to the very young, the elderly, and people with compromised immune systems (from long-term illness or exposure to radiation). Staphylococcal enterotoxin B was disseminated by planes over a 40- to 50-kilometer downwind grid. According to tracking records, a single weapon was calculated to have covered 2,400 square kilometers (926.5 square miles).

While the U.S. military made no effort to officially document the effects of this agent on the local Marshallese population, the Brookhaven medical survey did visit Rongelap before and after the test series and reported "a rather serious outbreak of Hong Kong influenza occurred among the Rongelap people in 1968 [that] may have been responsible for the deaths of a 58-year old exposed woman and of an unexposed boy who died of meningitis complicating the influenza."[4] Elsewhere, the March 1968 survey report noted that antibody tests for Asian influenza had been conducted, with negative results.

In the questioning that followed this testimony on the possibility that biochemical weapons testing had adversely affected the health of the Marshallese, Dr. Pollock

[4] Robert Conard, et al., *Medical Survey of the People of Rongelap and Utirik Islands Thirteen, Fourteen, and Fifteen Years after Exposure to Fallout Radiation (March 1967, March 1968, and March 1969)*, (Springfield, VA: Brookhaven National Laboratory, June 1970). On October 31, 2001, Councilwoman Rokko Laninbelik read the names of those in the Rongelap community who had died from the 1968 "influenza" epidemic into the Nuclear Claims Tribunal record. They were Martha Laudam, Tokjeta Riklon, and Rocky Job Jilej.

offered an impromptu and suggestively supportive story. In 1968 Dr. Pollock was conducting anthropological fieldwork on Namu Atoll in the southern Marshall Islands. At the time of the military exercise involving staphylococcal enterotoxin B, she was attending a baseball game with much of the island population. Partway through the game, everyone was hit with cramps. The game abruptly ended when everyone—athletes and fans—began vomiting. She later learned that there was a nationwide outbreak of severe flu. She described the symptoms as similar to food poisoning, with extremely rapid onset. Her personal experience of the islandwide outbreak on Namu suggests that the biochemical weapons tests held off Enewetak, like the earlier nuclear weapons test, might have had an adverse impact throughout the Marshall Islands.[5]

A third element that emerged during the hearing was a greater understanding that the Rongelap people not only experienced pain and suffering from their initial acute radiation exposure but also suffered from the abusive ways in which their injuries were documented, decontamination was attempted, and their broader pain and suffering was ignored. During the hearing, we posed a series of questions and heard lengthy testimony concerning the acute exposure from Bravo fallout and the experience of suffering radiation burns. At the break, a judge took Holly Barker aside to ask why we asked so many questions about this topic when the tribunal already provides compensation for radiation burns under the personal-injury program. What the judge had not realized years before, during the personal-injury phase of the tribunal process, and what was evident in new testimony at the hearing, was that people did not receive any pain medication or treatment for their burns. This hearing was different, because survivor recollections allowed judges to understand that radiation-burn injury involved more than physical suffering from burns that went bone-deep. Injuries also included the physical and psychological trauma

---

[5] The revelation that the Marshall Islands population experienced ill effects from biological weapons tests off their shores prompted formal bilateral requests from the RMI to the United States for a full and complete disclosure of all biological and chemical weapons tests conducted in Marshallese waters. For a news report on this issue, see, Giff Johnson, "Marshall Islands: Marshalls Chemical, Biological Tests Revealed," *Pacific Magazine,* June 11, 2002, http://www.pacificmagazine. net/news/2002/06/11/marshall-islands-marshalls-chemical-biological-tests-revealed (accessed April 1, 2008). Veterans Administration lobbying and congressional hearings prompted additional disclosures on Project Shad events and possible health effects on U.S. soldiers participating in these tests; a health care program for veterans of Project Shad was established by a 2003 act of Congress. A National Academy of Sciences study was funded to examine the long-term health effects of Project Shad exposure. Its May 2007 report, *Long-Term Health Effects of Participation in Project SHAD,* found "no clear evidence that specific long-term health effects are associated with participation in Project SHAD." William F. Page, Heather A. Young, and Harriet M. Crawford, *Long-Term Health Effects of Participation in Project SHAD* (Washington, DC: National Academies Press, 2007), http://www.iom.edu/?id=4909 (accessed February 10, 2008). Immediate effects from acute exposure, as suggested by the deaths of an older woman and a Rongelapese boy, were not explored. The Marshall Islands has yet to receive a formal response to its 2002 request.

associated with a medical response that documented injuries in careful, clinical terms but did not treat them. Testimony made clear that in the initial exams, and in the subsequent three months of medical care, U.S. government doctors and researchers measured and photographed people and their burns; took samples of blood, skin, urine, and feces; and supervised public decontamination scrubbings in the lagoon but, according to survivor memories, did not give any medication to treat injuries and relieve pain.[6]

A fourth major element emerging in the Nuclear Claims Tribunal hearing came from joint testimony offered by Almira Matayoshi and Lijon Eknilang, who recounted their memories of pain, inadequate response in terms of holistic medical treatment, and the enduring consequences of these experiences. While these issues are explored in some detail in the report, the hearing provided an opportunity to consider the consequences of this history. Testimony, especially by Almira Matayoshi, pointed to the continuing and escalating problems resulting from the colonial-era decision to divide health care obligations between two arms of the same government: radiogenic research on human populations and treatment for radiation-related cancers was the responsibility of the Atomic Energy Commission and then the Department of Energy; public health matters were the responsibility of the Pacific Trust Territory government, which was under the authority of and funded by the U.S. Navy and then the Department of Interior. When the

---

[6] The emphasis on study, rather than treatment is reflected in E. P. Cronkite et al., *Study of Response of Human Beings Accidentally Exposed to Significant Fallout Radiation* (Bethesda, MD: Naval Medical Research Institute; San Francisco: U.S. Naval Radiological Defense Laboratory, 1954). The plan for follow-up research is articulated in a letter from E.P. Cronkite, Naval Medical Research Institute, to John Bugher, Atomic Energy Commission, June 3, 1954, http://worf. eh.doe.gov/data/ihp1d/400132e.pdf (accessed April 1, 2008). The October 1954 report notes that in isolated cases, antibiotics were given to abate high fevers associated with an epidemic of respiratory illness but were generally withheld out of concern that use would reduce efficacy in the long run. For patients who developed severe burns, treatment was typically limited to topical application of calamine lotion with 1 percent phenol or, in a few cases, pontocaine ointment and other salves to soften skin. No reference can be found to the use of pain relievers. The issue of research versus treatment became a major source of conflict between Marshallese subjects and U.S. scientists over the four-decade life of the biomedical survey program. A 1961 exchange of memos between the public health officer and the high commissioner of the Pacific Trust Territory, regarding communications between Dr. Conard and the high commissioner and the desire to create a long-term agreement for continued research access to the exposed population (including the ability collect samples of fetus specimens), offers a clear sense that the medical survey program operated for the explicit purpose of gathering biological samples that allowed the objective study of radiation effects and not in any way for the medical treatment of the exposed Marshallese. See Dr. Robert Conard, memorandum to Dr. H. E. MacDonald, May 25, 1961. On file at the Embassy of the Republic of the Marshall Islands, Washington, DC; Dr. H. E. MacDonald, memorandum to the high commissioner, May 29, 1961. On file at the Embassy of the Republic of the Marshall Islands, Washington, DC. MacDonald writes, "All we ask is that they [the AEC] tell us specifically what they want and how they want it furnished to them. We will perform that service and they can do the research they desire in the laboratory."

Marshall Islands moved from a U.S. territorial possession to a freely associated independent state, the transition occurred without a viable holistic medical system to effectively monitor and treat a nation of people who had played host to U.S. nuclear war games.

We reprint here a translated portion of Almira Matayoshi's testimony, as it further illustrates the immense difficulties encountered by the Rongelap people as they struggle year in and year out to secure basic health care. In preparing this translation we realized that the video-transcript captures the body language, tenor of voice, and sense of anger and weariness that words on a written page fail to communicate. Thus, where possible, we include descriptive gestures.

First, Holly Barker asked: "What kind of medical programs do the people of Rongelap need today?" Lijon Eknilang, a woman serving as a witness alongside Almira Matayoshi, gave a long list of medical needs for illnesses that the United States argues are not linked to radiation and are therefore outside the scope of U.S.-funded medical programs. She listed "problems with eyes, ears, lumps that appear on the body, and women's illnesses in their stomachs [miscarriages]." After Lijon's response, Almira Matayoshi spoke:

May I also add a little more to explain? This is based on my experience in Hawaii with these two programs [DOE and 177]. . . . I always go to these programs to ask for help. Not only for me, but for my children, because in the 177 and DOE programs, your kids can't see doctors for radiation illnesses, and our medication always used to be free.

The times I have gone to Washington [D.C.] to ask for help, and they answer and say they will provide me with money to go to the doctor for every kind of illness. Then I come back [from Washington D.C.]. Now there is a new policy, and when I go to get services from 177 . . . today, when I go to get help, they say: "We don't have responsibility for those kinds of illnesses—medicine for diabetes and high blood pressure. We don't provide services for eyes, or fix teeth and put fillings in teeth. For all those things, go to DOE." Then I go to the other people, [Dr.] Neal [Palafox] and Dr. Preston [from DOE], and I tell them what I need directly to their faces, and they say: "Well, it will be okay. We'll do that. . . . Yes. We'll make improvements. We'll do all those things [forward hand motion denoting a list of items she requested].

Some people, including the people who stay in Hawaii and those who stay here in the country [RMI], it is like they [U.S. government medical providers] raise their hands and slap us over this way [slapping motion to her left], and then over to 177 they slap us from here to there [slapping motion to the right], and none of us know where we're supposed to get help from. . . . They tell us: "You go to 177 [pushing motion to the right] so they can take responsibility and give you your medications." Then we go there and they say: "We won't give you anything because we don't take responsibility for that [gestures to her left]. Go to DOE, because they'll give you what you need." And then we're told to go somewhere else. . . . We don't know where we're supposed to get permission from—help.

I have used my money to go to Washington and Congress to ask them to give me money for the doctors. Many times I have asked that they grant the authority to let me get assistance someplace. . . . No response up until today. I went to South Africa with women from countries all over the world and explained my dilemmas to them, the problems that I have. I flew to Tahiti. They asked me to go there for a meeting. Today, there's nothing. I haven't seen any results. . . . I'm told by people with responsibility that all our needs will be taken care of—each of my injuries since I've been affected [by the radiation]. . . . And I think these things need to be recognized for those whose lives have been damaged [states forcefully] and get them the medicine they need for their injuries. We need assistance for all our injuries because no one is sure what causes our problems. . . . I am asking you to help if you would please help.

I have to pay for my medication from the hospital all the time. It costs sixty dollars for my medication. And the medication for eyes [closes and rubs her eyes] I need because my eyes were burned from the fallout. Not a single thing in my body is good. And, again, I think that you won't fully recognize the needs of those who were injured and don't just tell us to go to 177 [repeats the slapping motion with her hand], or go to DOE [slaps her hand in the opposite direction], or go see someone else [waves her hand in a third and forward direction. Pauses].

There still is no help. I have fought so hard—many times I have—you know that I've been to Washington many times, have gone to Hawaii so many times. I've sat in front of you and spoke to you many times. I talked about my medication needs, my children's needs, and my injuries. . . . I've been told by [Dr.] Preston: "Get your insurance, and then we can attend to your needs, and we can give you your medication" [Pauses].

I don't have the ability to go to Japan anymore [waves her hand to the right], or go over here [waves hand to the left], or come to Hawaii [brings her hand in close]. Or to go to Washington and ask for assistance. They answer me and say, "Yes, we will give you your medication—the thing—money for the medical programs." When I return and go to 177 and go to DOE [motions to her right], or go over here [motions to the left], nothing gets done. But I pay for services. When I don't have money, I don't have medication. They say I really need to take the medicine for my high blood pressure to control my illness. If I don't have money—sometimes I try to give them five dollars and get five dollars worth of pills, but the cost for a prescription is sixty dollars. When I don't have money, I ask them to give me five dollars' worth, or ten dollars' worth. Sometimes I've gone and told them I don't have a cent, but I need my medication for high blood pressure, and when I ask them for five or ten dollars' worth of medicine, they don't give me anything because I don't have the sixty dollars.

I asked Congress to give me just a little money to go to the doctor, and they said: "We'll give it to you." Nothing happens for so many reasons. [Pauses and stares at the three judges] . . .

And these illnesses, where do they come from? Where will they show up in the body of a person?[7]

Over the course of three days, some fourteen and half hours of testimony fleshed out a history and chain of events and gave a small glimpse of the huge consequential damages incurred by the people of Rongelap as a result of the U.S. nuclear weapons testing program and the related opportunistic research program. For those in attendance—the Rongelap community, lawyers, expert witnesses, judges, and court officials—the hearing was, in a sense, "our Nuremberg tribunal," and more. The expert witness report and the tribunal hearings served as a truth and reconciliation commission, with Marshallese experts providing the testimony and the declassified narratives of scientists and scientific findings providing the damning substantiation. The participatory nature of the hearing allowed us all to engage in the struggle to understand the history of a people, their exposure to nuclear fallout, their struggles to survive radiogenic disease and sustain life in a contaminated environment, their difficulties in securing comprehensive medical treatment while serving as radiation research subjects, and the many problems of a people displaced from their traditional homeland. After decades of denial by the U.S. government, this was truly a historic hearing. Substantiated testimony was presented in a formal court setting and received with clarifying questions and re-spectful instances of shared emotion.[8] In defense of the claim, the expert witness for the Defender of the Fund presented a summary that further substantiated claimant concerns.[9]

At the end of the hearing, after the three judges, clerks, translators, press, and lawyers had left the room, the people of Rongelap announced to us and to each other their own conclusions: "We won! We do not need to hear the judges' findings, we know we have won." The women sang and danced, and we shared food and gifts. It was a rare honor to be in that room with these courageous, generous, and determined people who have struggled to have their story heard and acknowledged

---

[7] Video transcript, Nuclear Claims Tribunal, Nitijela Conference Room, Majuro, November 1, 2001.

[8] On three separate instances during this hearing, the testimony of Marshallese witnesses left many in the room, the judges included, in tears.

[9] In defense of the claim, on November 2, 2001, Dr. Nancy J. Pollock presented a methodology for quantifying appropriate compensatory damages, based on her analysis of the Marshallese diet, and suggestions for a three-tiered approach to assuming levels of exposure and thus consequential pain and suffering. Her testimony accepted the major framing of the chain of events and consequential damages outlined in our expert witness report and included several supportive references to our report and the case presented by the Public Advocate. She later published an elaborated version of her expert witness report: Nancy J. Pollack, "Reconstructing Diets for Compensation for Nuclear Testing in Rongelap, M. I.," in *Researching Food Habits: Methods and Problems*, ed. Helen MacBeth and Jeremy MacClancy (New York: Berghahn Books, 2004), 169–80.

by the U.S. government for more than fifty years. The power of these moments cannot be overstated.

Post Hearing

While the NCT hearing on the Rongelap claim had power to affect the moment, many years passed before the Nuclear Claims Tribunal rendered its judgment. In the intervening years, from 2001 to 2007, additional aspects of the claim were explored, new claims for other atolls were filed, and people carried on with their lives in a rapidly changing, increasingly militarized world.

As suggested by the nationwide outbreak of "Asian flu" following biological weapons tests in the northern Marshalls and as documented in the declassified military reports, the entire Marshall Islands were subject to fallout. Numerous atolls in the Marshall Islands—Kwajalein, Ujae, Likiep, and Ailuk, to name a few—have their own unique histories related to the testing program. The opportunity to file claims with the Nuclear Claims Tribunal, and the participatory and transparent nature of the claims proceedings, has prompted efforts to revisit and reclaim local histories. This broader public engagement, in turn, is allowing the nation to come to terms with the fact that damages and injuries resulting from the nuclear weapons testing program involve not just the four communities recognized by U.S. public law but the nation as a whole.

The people of Ailuk Atoll, for example, were not evacuated immediately after the 1954 Bravo incident, despite the fact that radiation levels on the atoll were high enough to warrant their removal. The rationale given for this decision by the U.S. Navy was logistical: it was simply too cumbersome to evacuate the 401 Ailuk residents from their home islands. As a result, the U.S. government decided to leave the people of Ailuk in their contaminated setting, without medical care and without information about how to reduce their exposure to contaminants from the Bravo test fallout or protect themselves from further exposure during subsequent tests in the 1954 series and in 1956 and 1958.[10]

The U.S. government's policy on nuclear damage and injuries to Marshallese communities reflects fixed notions of space and time. It assumes that dangerous levels of exposure to radiation occurred through one detonation in 1954, rather than through the entire series of sixty-seven atomic and thermonuclear tests. Its definition of the dangerous fallout zone encompasses only four atolls in the

---

[10] See R. A House, "Discussion of Off-Site Fallout," in *Operation Castle, Radiological Safety, Final Report*, vol. 1, *Spring 1954* (Washington, DC: Joint Task Force Seven, November 8, 1954). ACHRE remarked on this report: "Based on the initial reading of 20 roentgens, the U.S. task force should have evacuated the people of Ailuk. But according to Colonel House, 'the effort required to move the 400 inhabitants,' when weighed against potential health risks to the people of Ailuk, seemed too great, so 'it was decided not to evacuate the atoll.'" ACHRE, *Final Report*, chapter 12.

northern Marshall Islands, rather than the entire nation. And because current policy is derived from circa-1954 disclosures of contamination and understanding of risk, the U.S. government has effectively limited its liability. Only those people who were present on Rongelap and Utrik during the 1954 Bravo event are considered dangerously exposed (and thus eligible for U.S.-provided health care for radiation-related illnesses).[11]

U.S. government assumptions fail to recognize many other groups of people exposed to dangerous levels of contamination through their customary movement throughout the islands for child care, schooling, marriage, or employment. They fail to recognize groups of people who lived in less contaminated settings but consumed fish or crops harvested in heavily contaminated settings. For example, the DOE hired Marshallese workers from around the Trust Territory to cleanup the ground-zero locations of Bikini and Enewetak. These workers were exposed to considerable hazards and received little to no radiation safety information. Yet because they fall outside the temporal and spatial categories constructed by the U.S. government, they have historically been excluded from the health care monitoring and treatment programs provided by the United States for RMI populations exposed to radiation.[12]

---

[11] Republic of the Marshall Islands, *The Republic of the Marshall Islands Changed Circumstances Petition to Congress Regarding Changed Circumstances Arising from U.S. Nuclear Testing in the Marshall Islands, September 11, 2000* (Majuro, RMI: Republic of the Marshall Islands, 2000), http://www.bikiniatoll.com/petition.html (accessed February 11, 2008). The petition was revised on June 19, 2003; see http://www.yokwe.net/modules.php (accessed February 8, 2008). The RMI's petition to the U.S. Congress, based on new information, new international and U.S. safety and measurement standards, and NTC awards, requests funding to enable the NCT to make full payments of all claims, and for the medical infrastructure and services needed by populations affected by the U.S. nuclear weapons testing program. The Bush administration's response to the changed circumstances petition was released in November 2004. See Bureau of East Asian and Pacific Affairs, U.S. Department of State, *Report Evaluating the Request of the Government of the Republic of the Marshall Islands Presented to the Congress of the United States of America: Regarding Changed Circumstances Arising from U.S. Nuclear Testing in the Marshall Islands Pursuant to Article IX of the Nuclear Claims Settlement Approved by Congress in Public Law 99-239* (Washington, DC: U.S. State Department, November 2004), http://www.state.gov/p/eap/rls/rpt/40422.htm (accessed April 1, 2008). A subsequent Bush administration response was delivered to Congress on March 14, 2005. See Thomas Lum, Kenneth Thomas, C. Stephen Redhead, David Beardon, Mark Holt, and Salvatore Lazzori, *Republic of the Marshall Islands Changed Circumstances Petition to Congress* (Washington, DC: Library of Congress, March 14, 2005). For a detailed perspective on contamination, exposure, and remaining risk as an argument against the Marshall Islands petition, see Steven Simon, testimony the Senate Committee on Energy and Natural Resources, July 19, 2005. For perspectives and data in support of the petition, see James Plasman, Nuclear Claims Tribunal judge, testimony to the Senate Committee on Energy and Natural Resources, July 19, 2005, http://bulk.resource.org/gpo.gov/hearings/109s/24536.pdf (accessed February 10, 2008).
[12] See Peter Turcic, director of Division of Energy Employees, Occupational Illness Compensation, U.S. Department of Labor, Bulletin 02-30, September 12, 2002. Subject: Reissue—Suspension Code for Marshall Islands Cases, http://www.dol.gov/esa/regs/compliance/owcp/eeoicp/

These are just a few of the examples of communities whose injuries from the testing program were ignored and effectively silenced during the decades when the United States retained administrative control of the islands. As mentioned above, their histories are now beginning to emerge within the context of Nuclear Claims Tribunal proceedings. These hearings, and the NTC findings, challenge the official U.S. government account of the activities and consequences of the U.S. nuclear weapons testing program in the Marshall Islands.

What outsiders can never appreciate is how profoundly cancer and radiogenic illnesses have permeated the fabric of life in the Marshall Islands. Virtually every family in the Marshall Islands knows firsthand the struggle to find care for family members with cancer and the grief and loss that occur when loved ones succumb to the illness. Author Holly Barker lived on Mili Atoll when she was in the Peace Corps. This atoll is in the southeastern portion of the country and is the farthest place in the nation from the ground-zero locations. Barker's Marshallese mama, a woman who adopted her during her Peace Corps service, had a sister, Manuwe, who was married to a man employed by a DOE contractor to cleanup Bikini Atoll after the testing program. In the 1970s, Manuwe moved with her husband to Bikini, where she ate and drank from a contaminated environment. Manuwe died of cancer. But her death is not counted or considered part of the official burden experienced by the Marshallese, because she was from Mili, not from Bikini, Enewetak, Utrik, or Rongelap. Manuwe's husband is plagued by illnesses and is confined to a wheelchair. Because his exposure to radiation occurred while working on remediation projects, rather than the 1954 Bravo event, he is not eligible to participate in medical monitoring or treatment programs. Furthermore, because he is not a U.S. citizen, he is not eligible to participate in a U.S. Department of Labor compensation and medical care program for DOE workers exposed to radiation.

If this report is any indication of the regularity of cancer deaths in the Marshall Islands, we need look no further than the case of George Anjain, one of our advisors in the land-value study conducted in 1999. He passed away from an unexplained hemorrhage in the hospital on Ebeye (an 80-acre islet where some twelve thousand

PolicyandProcedures/finalbulletinshtml/bulletin02-30marshalisland-updated.htm (accessed February 11, 2008). As guidance for implementing this act, the Division of Energy Employees notes that if an injured worker was involved in DOE work in the Marshall Islands, a hold should be placed on the processing of compensation claims, especially for claims involving non-U.S. citizens, including Marshallese workers. In 2007 a bill was introduced in Congress to address this omission of coverage. As of this writing, the bill remains in committee. See S1756 Republic of the Marshall Islands Supplemental Nuclear Compensation Act of 2007, http://www.govtrack.us/congress/bill.xpd?bill=s110-1756 (accessed February 11, 2008). For a more detailed discussion of efforts to secure from the U.S. government a comprehensive resolution to nuclear legacy issues in the Marshall Islands, see Holly M. Barker, "From Analysis to Action: Efforts to Address the Nuclear Legacy in the Marshall Islands," in *Half-lives and Half-truths: Confronting the Radioactive Legacies of the Cold War,* ed. Barbara Rose Johnston (Santa Fe, NM: SAR Press, 2007), 213–47.

Marshallese live across the lagoon from a U.S. missile testing facility on Kwajalein Island). Anjain's death at the age of forty-nine was especially traumatic, as he was unable to receive needed emergency care because of a provision in U.S. law stating that the DOE cannot pay for any medical conditions considered "nonradiogenic."[13] Given Anjain's exposure to radioactive fallout from the Bravo test as a three-year-old child on Rongelap, and subsequent exposures due to life in an environment heavily contaminated by radiation, it is difficult, if not impossible, to establish a clear-cut distinction between the "direct" and "indirect" health consequences of the testing program.

Beyond George Anjain's death in 2000, we have lost the Marshallese judge on the Nuclear Claims Tribunal who had responsibility for the Rongelap land claim, Chairman Oscar DeBrum. DeBrum was a Marshallese ambassador and statesman and former district administrator for the Trust Territory government. His work had him traveling to all the contaminated atolls for meetings with communities or to monitor the progress of environmental remediation. Chairman DeBrum died in 2002 from prostate cancer.

In 2004 we lost John Anjain, who was mayor of Rongelap during the Bravo event. He was also a key spokesman for the community who carefully documented all the illnesses and change experienced by the Rongelapese in his private journals. John was George Anjain's father. He was also the father of Lekoj Anjain, a one-year-old baby at the time of Bravo and the first Marshallese to contract leukemia. Lekoj died at the age of nineteen. John died complaining of stomach pain. He had an undiagnosed cancer.

In 2005 we also lost Almira Matayoshi. She was an important matriarch who shared her experiences and the stories of the Rongelap people through formal testimony to the U.S. Congress, the United Nations, forums in Japan, and other venues around the world. She was a key witness during Rongelap's hearing before the tribunal. She testified about her personal experiences with exposure, evacuation, loss of land, medical exams, thyroid surgery, and reproductive health problems. She also discussed stereotypes imposed upon the Rongelapese and their ostracism in the broader Marshall Islands community by people who were fearful that they would catch radiation illnesses from the *ri-baam* (people of the bomb), before they understood the linkage between exposure and illness. Matayoshi died from bladder cancer.

---

[13] Program providers, as well as recipients, are frustrated by the severe limitations imposed by harsh and inappropriate eligibility language. For a description of the medical program flaws, inadequacies, and needs, see the critical comments and proactive suggestions offered by Neal Palafox, statement to U.S. Congress, July 19, 2005, http://www.yokwe.net/modules.php?op=m odload&name=News&file=article&sid=1139 (accessed February 10, 2008).

These are just a few of the many Marshallese involved in this research whose lives were constantly compromised by their struggles with radiogenic illness. The Marshallese have suffered more illness, death, and grief than any population should endure, and historical wrongs resulting from the nuclear weapons testing program have been compounded by inadequate and underfunded medical assistance. Despite the seriously elevated cancer rates in the Marshall Islands, as of this writing there is no oncologist in the country. There is no ability to provide chemotherapy or radiation treatment. Perhaps worst of all, there is no ability to undertake a nationwide screening for cancer to catch the illness in its early stages and provide patients with the greatest chance for survival and an improved quality of life.

The problematic consequences of the U.S. effort to address the health care needs of a radioactive nation were amplified for us in interviews conducted as follow-up research on the human radiation experimentation program. On September 30, 2004, Holly Barker interviewed a number of Rongelapese who had served as controls for the AEC-funded research conducted by Brookhaven Laboratory. Their experiences include an array of health issues. Because they were "controls" rather than "exposed," responsibility for their long-term health care is problematic. The Department of Energy has provided some medical attention over the years but in recent years has proposed terminating assistance to this population, to save money and because control subject health care is not specifically identified in congressional mandates. One of the resounding themes in these interviews is the frustration expressed by people who served as medical research subjects for so many years yet do not have access to their medical records, have immense health problems, do not have a full understanding of these problems, and are constantly troubled by the question of what ailments might indeed be a result of their radiogenic exposures. A few excerpts are reprinted here to provide some sense of the complex nature of serving as a control subject, as well as the difficulties in securing adequate health care. Because these interviews occurred outside of formal legal proceedings, to protect people's privacy we use their initials rather than their full names or human subject numbers.

T. A. was born on Jaluit in 1942. In 1967 he moved to Rongelap with his wife, a Rongelapese. One year later, he was enrolled in DOE's program as a control subject. No one asked his permission. He was simply told that he must participate. He understood it was for comparison purposes. He remembers x-rays, eye exams, unknown injections, and blood samples: "I was scared. I'm scared of doctors' equipment. I went to see if I had illnesses. I'm concerned about radiation [effects]." "I love my wife so I went to Likiep. [I participated in the exams because] I thought I was helping the exposed population." T. A. had an operation on a stomach tumor at Straub Hospital in Hawaii. He also has thyroid nodules but hasn't had thyroid surgery. He takes nine pills every day. He doesn't know what they are all for. He is worried about where he will get medical help if the U.S. government removes the control population from its current medical program.

J. K. was born in 1934 on Kosrae and lived on Ujae for the Bravo test. He was resettled on Rongelap with the exposed population in 1957. J. K. understood that the program was for comparison purposes. He thought it was a good thing to participate in the program since he was living on Rongelap, an irradiated place, and eating the local foods. J. K. remembers that U.S. government doctors gave him a blood transfusion and told him it would make him stronger. He had injections and never knew what they were for: "I was nervous and thought I wouldn't go [to the exams], but I always went." J. K. has been on thyroid medication since 2001. He has diabetes, high blood pressure, and eye problems. He has never been to the United States to see doctors.

A. J. was born on Rongelap in 1933 and was on Kwajalein during the Bravo test. He remembers, "The U.S. government told me to be a control. I went back in 1957. I did as I was told. They told me to." A. J. has swollen lumps on many parts of his body, including his breast, neck, arm, and behind his ears. He has had procedures to have them removed four times, but not in the United States. He also has breathing problems. He remembers that U.S. doctors gave him a blood transfusion. He doesn't know why. He did not get a letter from DOE stating he was part of the acknowledged human radiation experiments: "They've taken samples from me. . . . I didn't want them to enroll me, but they did. I went to Rongelap and I need to know I'm safe."

I. E. was born in 1940 on Rongelap and was on Ebeye during the 1954 Bravo test. The AEC asked for volunteers to compare their health to the exposed. "The people of Rongelap were my relatives. I was sad for them," I. E. recalls. "They had me hold up a number, and they took our picture. I had never heard of Project 4.1." "I met [Dr. Robert] Conard. He came twice a year. I don't know what they did to us. A Japanese doctor came with him once." I. E. remembers the blood samples and physical exams. He remembers receiving three kinds of injections, including one that he believed would help him be immune to illnesses. I. E. assisted the medical survey team: "They didn't explain things. They gave me number results that didn't make any sense." "They studied me. I might be poisoned, but they won't say. I ate things I shouldn't have, like coconut crab. I lived and ate where there is radiation. If I hadn't gone back to Rongelap, I wouldn't need this [medical] program." I. E.'s older sister, now living in Hawaii, received a letter from the Department of Energy saying that she had been a human subject in experiments involving chromium-51 and tritiated water.

K. A. is from Ujae. She was born in 1943 and moved to Rongelap in 1957. She said that U.S. doctors asked for volunteers for blood comparisons: "I was helping them [the exposed]. We were sad for them. . . . [U.S. government scientists] would give us candy and fruit and milk to be in the program. They really wanted us then." She remembers that the lobster, pandanus and fish she ate on Rongelap caused swollen bumps in her mouth. She remembers blood samples, stomach exams, and experiencing swollen legs. She had injections but doesn't know for what. She says she has heart problems. On two different occasions, the U.S. doctors cut open her side and put tubes in her. She received no papers describing the procedures. She complains of severe body pain, especially in her legs. At times she cannot walk: "I have pain constantly. I need a doctor." "My eye tears incessantly. . . . They cut something hard out from under my eye." "My heart's getting bigger—I have a bad problem." U.S. government doctors told her that she had breast cancer but then told her that she was fine (benign tumors?). She says, "I want a doctor for my health. Rongelap hurt my body. I still have skin problems from Rongelap. I still have eye problems. My teeth fell out on their own from the poisonous food on Rongelap. My parents died young." K. A.'s mother had breast cancer. Her father had swollen legs. Her sister died of bone cancer. All three of these family members were also control subjects.

What meaningful recourse do these people have, given the immense nature of their problems and complaints? A minimalist approach to health care has been provided through the Compact of Free Association (177 Agreement): Some seventeen thousand people receive health care through the 177 Health Care Program established to address the radiogenic health issues of the people of Enewetak, Bikini, Rongelap, and Utrik islands. This system is woefully underfunded and lacks comprehensive cancer treatment capability. As the testimony of Almira Matayoshi suggests, many people have filed personal-injury claims and, with their compensation, moved to Hawaii and the continental United States seeking, among other things, better health care. The NCT has ordered millions of dollars in compensation for personal-injury claims, but many more people have been found eligible than originally anticipated. Thus the majority of awards have yet to be paid in full to victims or their surviving families.[14] And while a compensatory payment provides assistance at one level or another, in no way does it provide the means to restore overall health. What is clearly lacking in the Marshall Islands, and

---

[14] The 177 Health Care Program includes people from Bikini, Enewetak, Rongelap, and Utrik deemed by the United States to be exposed; thousands of other residents of the four atolls whose coverage was deemed eligible by local atoll governments in the first fifteen years of the 177 Agreement program; and other people who received NCT personal injury awards. The Nuclear Claims Tribunal personal injury program currently recognizes some thirty-six cancers and radiogenic health conditions. By December 2006, more than seven thousand claims had been filed and some two thousand Marshallese had been deemed eligible for personal injury awards. Many received awards for multiple ailments. Most awards (1,186) were for thyroid cancers and disease, pulmonary and lung cancer (235), cancers of the blood, bone marrow, and lymph nodes (143), breast cancer (111), and cancers of the ovary (61). As of this writing, no personal injury claim has been paid in full. The last annual pro rata payments, made in October 2005, brought the cumulative payout for those whose awards were issued prior to October 1996 to 91 percent. Awards issued after that date have received varying levels of payment, ranging from 5 to 84 percent (based on Barbara Rose Johnston's review of Nuclear Claims Tribunal award records and personal communication from Bill Graham, February 12, 2007).

sorely needed, is a high-quality medical care program that would address direct *and* indirect health problems caused by U.S. activities during the nuclear test period, and build the capacity of the Marshall Islands to address these needs.[15]

Seeking Meaningful Remedy

The Nuclear Claims Tribunal was forced to halt incremental payments for personal-injury awards in 2006 due to a lack of funds. As noted above, there have been far more claims for radiogenic illnesses in the Marshall Islands than anyone expected. And the cost to restore the damage created by the nuclear weapons testing program greatly exceeds any initial expectation, especially since the award fund was established during negotiations between the United States and its territorial possession, at a time when contaminant levels were still classified and kept secret. With the release of previously classified materials, and scientific research accompanying the development of personal-injury, property-damage, and hardship claims to the Nuclear Claims Tribunal, a more thorough understanding of physical injury, property damage, and costs to restore has been achieved.

In September 2000, the RMI government presented the U.S. Congress with a changed circumstances petition requesting that the U.S. government fund and provide adequate health care for all populations in the RMI whose health had been adversely affected by the U.S. nuclear weapons testing program, and fully fund the Nuclear Claims Tribunal so that it could make awards for personal injury and private property damage in accordance with its congressional mandate. "Changed circumstances" refers to any new information not known when the original Compact of Free Association was consummated and signed on June 25, 1983. New information includes not only public awareness that nuclear weapons tests adversely affected the entire nation but also new scientific evidence that low-level exposures to radiation produce significant health risks, and that these risks were not understood when the compact was originally negotiated.[16]

In 2004 the U.S. National Cancer Institute (NCI) predicted that more than five hundred cancers will ultimately manifest in the population that was living in the Marshall Islands in the 1950s as a direct result of the testing program. These are cancers that would not exist had the U.S. government not used the Marshall Islands to conduct sixty-seven atmospheric weapons tests. Because of the long latency period of certain types of cancer and the aging of the population, more than two hundred of those cancers have yet to surface. And 297 of the excess cancers

---

[15] See note 12. Palafox argues that it is the responsibility of the United States to provide the U.S. standard of care for cancer patients in the RMI because the U.S. nuclear weapons testing program caused the excess cancers.

[16] R. A. House, "Discussion of Off-Site Fallout," in *Operation Castle, Radiological Safety, Final Report,* vol. 1, *Spring 1954* (Washington, DC: Joint Task Force Seven, November 8, 1954).

estimated by the NCI will occur among people who in 1954 were living at atolls other than Rongelap and Utrik, the only atoll populations eligible to receive U.S. health care for radiogenic illnesses.[17]

In July 2005, the National Academy of Sciences released the *Biological Effects of Ionizing Radiation VII Report* (BEIR VII), reaffirming the conclusion of the 1990 BEIR V report that every exposure to radiation produces a corresponding increase in cancer risk.[18] Despite scientific evidence that additional information has come to light and that damages are far greater than originally anticipated, the changed circumstances petition has not been acted upon in Congress, in large part due to the Bush administration's review of the petition and the contentious finding that the U.S. government has met its' legal obligation to the Marshallese. At congressional hearings about the petition in July 2005, and again in 2007, the executive branch made it clear that it does not find merit in any of the RMI's requests, and Congress still lacks the political will to provide meaningful relief to the Marshallese.

Prior to the Rongelap hearing in 2001, the tribunal had completed its review and issued property-damage claim judgments, awarding hundreds of millions of dollars to compensate, remediate, and restore Bikini and Enewetak. In the years since, the tribunal has issued judgments in the Utrik claim and accepted other atoll claims.[19]

On April 17, 2007, some sixteen years after the first claims were filed, the Nuclear Claims Tribunal finally issued its decision in the Rongelap case. As laid out in the thirty-four-page judgment: "The Tribunal has determined the amount of compensation due to the Claimants in this case is $1,031,231,200. This amount includes $212,000,000 for remediation and restoration of Rongelap and Rongerik Atolls. This award further includes $784,500,000 for past and future lost property value of Rongelap, Rongerik and Ailinginae Atolls as a result of the Nuclear Testing Program. Finally, it includes $34,731,200 to the Claimants for consequential damages."

Notably, the "past and future lost property" award reflects "loss of way of life damages," including the loss of the means to live in a healthy fashion on the land; people were on the island but were exposed to high levels of radiation. And the consequential damages award not only includes the resulting pain, suffering, and

---

[17] National Cancer Institute, *Estimation of the Baseline Number of Cancers among Marshallese and the Number of Cancers Attributable to Exposure to Fallout from Nuclear Weapons Testing Conducted in the Marshall Islands* (Washington, DC: Division of Cancer Epidemiology and Genetics, National Institutes of Health, 2004).

[18] See National Academy of Sciences, *BEIR VII: Health Risks from Exposures to Low Levels of Ionizing Radiation* (Washington, DC: National Academy Press, 2005).

[19] Full text of the decisions for the Bikini, Enewetak, Utrik, and Rongelap awards, as well as details on the personal injury award program, can be found on the RMI Nuclear Claims Tribunal site, http://www.nuclearclaimstribunal.com/award.htm.

hardships from "loss of a healthy way of life" but also awards personal injury to subjects identified as receiving chromium-51 injections, which were "an additional burden to the already considerable exposure from consuming contaminated foods and living in a radioactive environment."

With regard to the larger involvement of the Rongelap people in four decades of human subject research, the tribunal found that "the emotional distress resulting from the participation in these studies and the manner in which they were carried out, warrants compensation, and is a component in the consequential damages related to the period of time the people spent on Rongelap from 1957 to 1985."[20]

It was with mixed emotions that we received the news of the final judgment in the Rongelap claim. Historical wrongs have been acknowledged. Yet the initial trust fund was never adequate to pay personal-injury and property-damage awards, and an act of the U.S. Congress is required before tribunal judgments can be implemented in full. By 2007 there was less than $1 million left in the NCT fund. The Rongelap award, prior awards to Bikini, Enewetak, and Utrik, and a huge portion of the personal-injury awards remain unpaid. Pending claims for other atolls can look forward to a similar status. At this writing, there is no political will in the U.S. government to right these wrongs. Whether or not the ordered remedy will ever be achieved remains to be seen.

Political administrations come and go, but radiogenic contamination and disease present protracted, ulcerating, intergenerational problems. The toxic and radioactive contamination of soil, water, terrestrial and marine biota, and human life that is the legacy of nuclear war games in the Marshall Islands is difficult and expensive to monitor, let alone remediate. The health complications of radiation exposure for individuals and their offspring are similarly expensive to monitor and treat. Nevertheless, it is our contention that just as the U.S. government continues to appropriate billions of dollars for the cleanup of the plutonium processing plant in Hanford, Washington, and as it continues to make appropriations to provide full compensation to people living downwind from the Nevada Test Site, so too must it honor commitments to the inhabitants of the former trust territory, who deserve the same level of health care and cleanup as U.S. citizens.

While actual payment on the Nuclear Claims Tribunal judgments remains in question, and thus the costly work of environmental remediation is in doubt, some of the recommendations for remedy outlined in *The Rongelap Report*—ideas that originated from within the Marshallese community—have spurred action in the Marshall Islands, thanks to the many efforts of Rongelap citizens and their government. When we presented *The Rongelap Report* as supporting evidence in the

---

[20] Full text of the Nuclear Claims Tribunal judgment in the Rongelap claim can be found on the NCT website, http://www.nuclearclaimstribunal.com/rongelapfin.htm#_ftn76.

Rongelap property-damage claim in 2001, it was a time of intense effort by the government of the Republic of the Marshall Islands to urge the government of the United States to take greater responsibility for damages and injuries linked to the testing program. It was also a time of purposeful effort by the people of Rongelap to define themselves not as victims, who are typically portrayed as incapacitated, passive, and unable to take control of their situation, but as survivors—people who experienced extreme hardship yet are able to take the active steps necessary to reclaim their futures. In the years since, collaborative research projects have been initiated by the Rongelap local government, with professors and students at the University of Hawaii producing culturally appropriate designs and structures that have since been built on Rongelap Island. A market analysis was conducted, and the local government purchased a boat, built tourism structures on Rongelap Island, and launched an ecotourism venture.[21] Plans were announced, and a fund-raising effort to build a Rongelap Peace Museum is under way.[22] The Rongelap local government hired a team of independent lawyers, health physicists, other scientists, and community planners to review data produced by the U.S. government about the future habitability of Rongelap and to review data by the local government's own researchers.[23]

Both U.S. scientists and Rongelap consultants believe that the island of Rongelap could be habitable year-round in the future if remediation of the soil continues. They also agree that other islands in Rongelap Atoll, especially the northern islands, will remain dangerously contaminated for many years to come (in some cases, twenty-five thousand years). Thus the question of whether the people of Rongelap will actually be able to return and live in a sustainable fashion on one relatively "safe" island within a still intensely contaminated atoll has yet to be resolved.[24] Will people survive and thrive when the exploitation of critical resources is largely limited to the main island of Rongelap? What measures can be taken to ensure that

---

[21] See, for example, the summary of projects and status of the resettlement program published on the Rongelap local government website, http://www.visitrongelap.com. The media center Web page includes links to newspaper and magazine articles on the status of restoring the Rongelap economy, environment, and way of life.

[22] A plan has been completed to build a museum on Majuro that would showcase the experiences of Rongelap and other radiation-affected communities. Funds are being solicited from other nations and the international community via a website. As of this writing, the plan is still in the proposal stage.

[23] For a summary of work conducted by Bernd Franke and the Institute for Energy and Environmental Research, see the publications list at http://www.ieer.org/pubs/index.html and Christian Liedtke, "Rongelap's Dream Becomes a Reality," *Marshall Islands Journal,* March 31, 2006, http://www.visitrongelap.com/MediaCenter/Press_Releases/pdf/2006_mar_resettlement.pdf (accessed February 11, 2008).

[24] The status of Rongelap remediation and resettlement efforts are also reported on the Department of Energy's Marshall Islands Dose Assessment and Radioecology Program website. See https://eed.llnl.gov/mi/rongelap.php (accessed February 11, 2008).

the people, as they grow and eat local food and live in their environment, are not at risk from hazardous levels of radiation? If the resettlement costs more than initially planned, who will be accountable? Will the Rongelapese feel comfortable moving back to their home islands a third time, given their experiences of returning in the past with false assurances that all was well? And what of the generations to come? Will the younger generation, born and raised in the more developed islands of the southern atolls, have an interest in returning to simpler life on an outer island? Do they have an affinity to a place where they were not born and raised? Will they view Rongelap as a place to live, realize their dreams, and raise future generations?

These difficult and complex questions can be addressed only when the community has full access to scientific information; has the means and ability to fully participate in developing and choosing remedial strategies and resettlement plans; and, because of its increased knowledge and ability to participate, can feel confident that its decisions rest upon transparent and verifiable science. Fortunately, the efforts taken by community leaders to implement recommendations in *The Rongelap Report,* conduct their own independent reviews, and thoroughly and regularly communicate their progress in the scientific assessment and resettlement planning process have greatly empowered the people of Rongelap. The struggle continues, but the people of Rongelap are well on the road to coming to terms with the realities of life in a radioactive nation.

Appendix

Sample Marshallese text from the memoir of John Anjain.

Supporting Documents. List of documents submitted to the Nuclear Claims Tribunal in support of the Rongelap claim.

Letter from the Advisory Committee on Biology and Medicine to Lewis Strauss, chairman of the U.S. Atomic Energy Commission, November 19, 1956.

Memorandum from Gordon M. Dunning to C. L. Dunham, June 13, 1957. Subject: Resurvey of Rongelap Atoll.

Letter from Hermann Lisco, MD, Cancer Research Institute, New England Deaconess Hospital, to George Darling, Director, Atomic Bomb Casualty Commission, April 29, 1966.

Letter from Paul Seligman, U.S. Department of Energy, to Mayor James Matayoshi, Rongelap Atoll Local Government Council, April 29, 1999.

Sample Marshallese text from the
memoir of John Anjain

Ekwe, bōtab ilo year ko lok emōtlok
3 im 4 year kiō, ewōr juon unok eo
Doctor ro An A.E.C. endwōj air loe im bōktok
Boktok, etan unok in
Uno in ej kein jolok eboj ilo burnōn Armij
Ilo tōmak eo aō, uno in ekanoj in lap an
Komōn bwe armij ro en mohōj air ejimur,
Im ebar einwōt komōn bwe Ajiri ro ren nitto
Mōkōj, im ij tōmak bwe enaj komōn, army roilo
Kan kane rij itok ren, kejur im bōjibōn
Uno in ejerar na ilo jikin ejimur ko ilo rellap
Ilo ailing kein, im ij limnōk bwe jabrewōt
Eo enaj lōn eboj ilo burnōn, enaj marlōn in
Mour jen uno in, jekron jen lea im jekron re
kar jab jorren, Ak ij tōmak bwe ilo ran kane
Rej itok elōn ren keyerbal uno in, im bōlen
Elōn rij keyerbal kiiō, ak ian armij ro rar
Jorren, elōn ro rej jerwane uno in im rej
Jab kanoj lomnōk kake, ekwe kain army rot in
Elap an bwir, im yab lomnōk ke mour.
Eo an yab an armij ro jet, ro me rej lalorjake
Ejimur non jabrewōt Army, barainwōt
Elōn wōt ro rej ebwer kin air bōk bitōktōkier
Bwe rej ba enanin lap bitōktōk eo rej bōk jen
Kij, ekwe, Na, ij ejab ein aō lomnōk wōt
Army rein, ak ne rej bōk bitōktōk jen na
Ej juon men eo elap an itok limaō kake.
Inem ikonan rejan aolep armij ro rej mour
Jen Rongelap ranin, ilo Neota bebartin iar jab konono
kij bariu, bwe enaj wōr juon bwebwenato kake.
Ekwe, Na John Anjain iar bōk jerbal in
Dri tel ilo Rongelap iomin 12 year ko iar
Dri tel non armij ro ilo Rongelap, im aolep
Men kein iar butbwenato kaki, iar loi im ronjaki
Im ij tōmak bwe elap lok aō melele im jela kake
Jen aolep dri tel ro rar julak tok elik.

John Anjain

Supporting Documents: List of documents submitted to the Nuclear Claims Tribunal in support of the Rongelap claim

To further illustrate points made and conclusions reached in our expert witness report, we submitted the following documents to the RMI Nuclear Claims Tribunal as an attachment to the report, as exhibits during the NCT hearing, and in support of post-hearing briefs. Web citations are given where documents are now publicly accessible. Those documents that are not easily accessed are reprinted here.

1. James Forrestal, letter to Gordon Sproul, June 2, 1947. The letter establishes a program of ecological studies documenting the movement of radiation through the atoll ecosystem in the Marshall Islands. It notes the presence of "radioactive trace substances in relatively large amounts." http://worf.eh.doe.gov/data/ihp1c/8665_.pdf
2. C. E. Wilson, memorandum for the secretary of defense to the secretary of the army, secretary of the navy, and secretary of the air force, February 26, 1953. Subject: Use of Human Volunteers in Experimental Research. See pages 20–25. http://www.defenselink.mil/pubs/dodhre/Narratv.pdf
3. Joint Task Force Seven, memorandum for the record. Subject: Bravo Shot, Operation Castle, April 12, 1954. See, especially, "Command Briefing, 0000, 1 March 1954, Submitted by R. A. House on 5 March 1954." http://worf.eh.doe.gov/data/ihp1c/0804_a.pdf
4. Colonel Clinton S. Maupin, staff surgeon, memo to commander, Joint Task Force Seven, April 10, 1954. Subject: Medical Conference at Kwajalein. Page 2 states, "In view of the fact that this group received a dose of radiation which was marginal from a standpoint of severe morbidity, justification cannot be made for exposure to significant additional radiation. Therefore, based on the concept that the recovery period should correspond in time to the permissible dose for accumulation, it is recommended that these patients not be exposed to radiation except for essential diagnostic or therapeutic radiation for a period of eight years." http://worf.eh.doe.gov/data/ihp1a/3259_.pdf
5. Division of Biology and Medicine, Atomic Energy Commission, transcripts, Conference on Long Term Surveys and Studies of Marshall Islands, July 12–13, 1954. Transcripts include detailed plans to conduct additional research on the movement of radioisotopes through the food chain and the Rongelap people, assuming a return to a contaminated setting. The EDTA experiment is also discussed as interesting although ineffective. See pages 54–55. http://worf.eh.doe.gov/data/ihp1c/0246_a.pdf
6. S. H. Cohn, R. W. Rinehart, J. K. Gong, J. S. Robertson, W. L. Milne, W. H. Chapman, and V. P. Bond, *Report to the Scientific Director. Project 4.1—Addendum Report. Nature and Extent of Internal Radioactive Contamination of Human Beings,*

Plants, and Animals Exposed to Fallout. The report includes description of an EDTA (ethylene diamine tetra-acetic acid) experiment involving seven Rongelapese between the ages of nineteen and sixty. http://worf.eh.doe.gov/data/ihp1d/6205e.pdf

7. Staff of the Applied Fisheries Laboratory, *Radiobiological Resurvey of Rongelap and Ailinginae Atolls Marshall Islands October–November 1955* (Seattle: University of Washington Laboratory of Radiation Biology, 1955). http://worf.eh.doe.gov/data/ihp1c/0696_a.pdf

8. Advisory Committee on Biology and Medicine, minutes, fifty-eighth meeting, Brookhaven National Laboratory, November 16–17, 1956. See pages 9 and 10 establishing a policy of no re-evacuation. http://worf.eh.doe.gov/data/ihp1d/1751_f.pdf

9. Advisory Committee on Biology and Medicine, letter to Lewis Strauss, chairman of the U.S. Atomic Energy Commission, November 19, 1956, regarding resolution establishing a policy of no re-evacuation, unanimously approved by committee members. Document originally accessed from the HREX site. Reprinted here.

10. Gordon M. Dunning, memorandum to C. L. Dunham, June 13, 1957. Subject: Resurvey of Rongelap Atoll. Complaint over resettlement without a complete survey of the atoll, "especially the foodstuffs." Document originally accessed from the HREX site. Reprinted here.

11. Gordon M. Dunning, memorandum to Dr. A. H. Seymour, February 13, 1958. Subject: Operational Responsibilities. http://worf.eh.doe.gov/data/ihp1d/400209e.pdf

12. Robert A. Conard, et al., *March 1957 Medical Survey of Rongelap and Utrik People Three Years after Exposure to Radioactive Fallout* (Upton, NY: Brookhaven National Laboratory, 1958). http://worf.eh.doe.gov/data/ihp1a/1024_.pdf

13. Robert A. Conard, letter to Charles L. Dunham, June 5, 1958. Letter concerns complaints of too many exams and no treatment, and problems with the ban on consuming crabs. http://worf.eh.doe.gov/data/ihp1d/400212e.pdf

14. Robert A. Conard, letter to Charles Dunham, January 16, 1959, requesting change in the policy forbidding studies of radioisotopes involving the Rongelap population. Describes three specific radioisotope studies. http://worf.eh.doe.gov/data/ihp1b/7783_.pdf

15. Charles Dunham, letter to Robert Conard, January 27, 1959, suggesting "no necessary relation between thyroid uptake and high protein bound iodine" and no need to "inconvenience" the people; approving plans for Cr-51 studies; suggesting radioisotope studies are limited to adults. See also handwritten note from Robert Conard to Dr. Neil Borss, February 1995, suggesting that only the experiment described in item 2 (involving the administration of Cr-51 to ten "nonexposed" Marshallese) was actually conducted. Note that Conard's 1995

recollection does not reflect data contained in his medical survey reports and related communications, which indicate that Cr-51, as well as tritiated water (tritium), was given to a much larger pool of subjects and that iodine-uptake studies were conducted with the administration of I-131, I-128 and I-129. http://worf.eh.doe.gov/data/ihp1b/7783_.pdf

16. Robert A. Conard, *An Outline of Some of the Highlights of the Medical Survey of the Marshallese Carried out in February–March 1959, 5 Years After Fallout.* Response to Rongelap community complaints of food poisoning and other health problems from radiation. The trip report references miscarriages and stillbirths among the Rongelap people. http://worf.eh.doe.gov/data/ihp1a/2698_.pdf

17. Robert A. Conard et al., *Medical Survey of Rongelap People, March 1958, Four Years after Exposure to Fallout* (Upton, NY: Brookhaven National Laboratory, Associated Universities, Inc., 1959). http://worf.eh.doe.gov/data/ihp1b/3543_.pdf

18. Robert Conard, letter to Courts Oulahan, AEC Deputy General Counsel, April 17, 1961. Response to Rongelap community complaints of food poisoning and other health problems from radiation. http://worf.eh.doe.gov/data/ihp1b/3813_.pdf

19. C. L. Dunham, letter to James T. Ramey, May 2, 1961, reporting "low body burdens of Sr-90, Cs-137 and Zn-65, all of which we believe to originate from the contamination in their current food supplies; the Zn-65 is believed to come from the seafood caught locally." http://worf.eh.doe.gov/data/ihp1a/3285_.pdf

20. Robert A. Conard, *Summary Report of 1959 and 1960 Medical Survey* (Washington, DC: Atomic Energy Commission, 1961). http://worf.eh.doe.gov/data/ihp1a/3285_.pdf

21. Robert A Conard, Leo M. Meyer, Wataru W. Sutow, William C. Moloney, Austin Lowrey, A. Hicking, and Ezra Riklon, *Medical Survey of Rongelap People Eight Years after Exposure to Fallout* (Upton, NY: Brookhaven National Laboratory, 1963). This survey includes findings from the administration of Cr-51 to thirteen Marshallese subjects and ten "Caucasian" subjects in 1961 and 1962. http://worf.eh.doe.gov/data/ihp1a/2682_.pdf

22. Robert A. Conard, et al., *Medical Survey of the People of Rongelap and Utirik Islands Nine and Ten Years after Exposure to Fallout Radiation (March 1963 and March 1964)* (Upton, NY: Brookhaven National Laboratory). This survey includes findings from the administration of tritiated water and Cr-51 to twenty-one Marshallese subjects. See page 39: "In order to establish the relationship of blood volume to lean body mass tritiated water was administered orally to each of 21 Marshallese subjects during the 1963 survey. In addition, determinations were made of red cell mass and blood volume by using Cr51-labeled sodium chromate." http://worf.eh.doe.gov/data/ihp1b/3547_.pdf

23. Leo M. Meyer, letter to W. Siri, April 9, 1965. Subject: Sending you the 1.0 ml volumetric pipette which we used at Kwajalein. Data from analysis of samples collected from Rongelap subjects in radioisotope studies reporting findings for six Marshallese subjects who received Cr-51 and tritiated water. http://worf.eh.doe.gov/data/ihp2/2522_.pdf

24. Robert A. Conard, et al., *Medical Survey of the People of Rongelap and Utirik Islands Eleven and Twelve Years after Exposure to Fallout Radiation (March 1965 and March 1966)* (Upton, NY: Brookhaven National Laboratory). The report notes: "Previous studies (1961, 1962) with Cr51-labeled erythrocytes on Marshallese subjects living in their native environment have shown reduced red cell mass and/or total blood volume with total body weight used as a base line. During the 1963 survey, similar studies were performed on 21 Marshall Islanders, but these data were related to total body water as determined by tritiated water. The present study was undertaken during the surveys in 1965 and 1966. A total of 19 Caucasian Americans (3 females and 16 males) living in the Marshall Islands for periods of 3 months to 9 years were examined by the same techniques. The results of these studies on each individual are presented in Appendix 15, along with data on the 21 Marshallese in whom these studies were carried out in 1963." http://worf.eh.doe.gov/data/ihp1a/2683_.pdf

25. Hermann Lisco, letter to George Darling, April 29, 1966. The letter reports findings on chromosome abnormality studies using samples collected by R. A. Conard in 1964. Notes findings reported in Hermann Lisco and Robert A. Conard, "Chromosome Studies on Marshall Islanders Exposed to Fallout Radiation," *Science* 157 (1967): 445–47. Document originally accessed from the HREX site. Reprinted here.

26. Robert A. Conard, *Protocol for the 1974 Medical Survey in the Marshall Islands.* Examples of human subject exams, procedures, and sampling undertaken to support a wide range of scientific research, including a thyroid reserve study and dietary iodine and iodine excretion level studies. http://worf.eh.doe.gov/data/ihp1c/0764_a.pdf

27. Robert Conard, letter to Dr. James L. Liverman, May 8, 1974. Reports outcome of the 1974 research and a summary of a Rongelap village meeting with continued "misunderstandings." http://worf.eh.doe.gov/data/ihp2/2829_.pdf

28. Tommy F. McGraw, Department of Energy, memorandum to Edward J. Vallario regarding DOE involvement in the evacuation of Rongelap Atoll. The memo illustrates that some U.S. scientists acknowledged the need for the Rongelap community to move in 1985 and acknowledged that continued occupation did indeed pose significant health risks. See also the eleven attachments that accompany this memo. http://worf.eh.doe.gov/data/ihp1d/400171e.pdf

29. U.S. Department of Energy, disclosure letter to Mayor James Matayoshi, April 29, 1999, regarding Cr-51 and tritiated water radioisotope studies. Reprinted here.

Letter from the Advisory Committee on Biology and Medicine to Lewis Strauss, chairman of the U.S. Atomic Energy Commission, November 19, 1956

UNITED STATES
ATOMIC ENERGY COMMISSION
NEW YORK OPERATIONS OFFICE
70 COLUMBUS AVENUE
NEW YORK 23, NEW YORK

709132

TELEPHONE NO.:
PLaza 7-3600

Refer to:
ACBM

November 19, 1956

Honorable Lewis L. Strauss
Chairman, U. S. Atomic Energy Commission
1901 Constitution Avenue, N. W.
Washington 25, D. C.

Dear Mr. Strauss:

At the 57th regular meeting of the AEC Advisory Committee for Biology and Medicine held at the Brookhaven National Laboratory on November 16 and 17, 1956, the following three motions were introduced and unanimously approved by the Committee members. Because of their urgency, it was agreed that they be transmitted to you immediately.

1. The ACBM reviewed with the staff of the Division of Biology and Medicine not only the Division's traditional duties, but also those many special additional responsibilities demanded by the present situation not envisaged eighteen months ago when the present personnel ceiling and budget were prepared. The Committee believes that under present circumstances, the Division cannot continue to deliver the present high standard of guidance of health policies and the provision of factual data without immediate reinforcement.

We note with satisfaction that this situation has been recognized for Fiscal Year 1958 and that in the last few days additional funds have been assured for 1958. However, the present situation as to funds, personnel and space, has been rendered acute by the emergent demands placed on the Division by its crucial role in determining the long-range effects of radioactive fallout and related matters. The immediacy of these requests requires action now.

1060052

Honorable Lewis L. Strauss   - 2 -   November 19, 1956

2. The Committee has reviewed Staff Paper AEC 141/33, the summary of which is as follows:

> "The trend in thinking of the International Commission on Radiation Protection, the National Committee on Radiation Protection, and the National Acadamy of Sciences is toward more restrictive criteria for standards of radiation protection.
>
> In light of the above, it is recommended that radiological safety criteria for exposures to gamma radiation from fallout to populations around the Nevada Test Site should be as follows:
>
> a. The current criterion of 3.9 roentgens for any one year; plus an additional restriction of
>
> b. 10 roentgens in a period of 10 years, with the first of the successive ten-year periods starting in the spring of 1951.
>
> These should be construed to be operational guides rather than maximum permissible limits, since exposures somewhat in excess would not be hazardous."

The above summary is approved in view of the Committee's opinion as to the necessity for continued weapons tests.

3. It is moved that the ACBM approve the Division of Biology and Medicine's proposal to return the Rongelapese to their native atol. However, it is the opinion of the ACBM that if it should become necessary to re-evacuate because of further tests, there would result world opinion unfavorable to the continuation of weapons testing.

The usual letter summarizing the entire meeting will be transmitted to you shortly.

Sincerely yours,

G. Failla, Chairman
Advisory Committee for Biology and Medicine

Memorandum from Gordon M. Dunning to C. L. Dunham, June 13, 1957. Subject: Resurvey of Rongelap Atoll

OFFICIAL USE ONLY

C. L. Dunham, M. D., Director   June 13, 1957
Division of Biology and Medicine

Gordon M. Dunning, Health Physicist
Division of Biology and Medicine

RESURVEY OF RONGELAP ATOLL

I had assumed there would be another radiological survey of the
Rongelap Atoll just prior to the return of the Rongelapese. I
learned from Dr. Seymour today that Dr. Donaldson's group would
not be returning until July for their next survey (the Rongelapese
will be returned the latter part of June). I felt it was quite
essential for documentary and public relations reasons that a survey should be made prior to, or at least at the same time as the
return of the Rongelapese. Therefore, I checked with Colonel
Schnittke and he is making arrangements for the RAD Safe group at
Eniwetok to make external gamma measurements. It would have been
highly preferable to have had a complete survey of the Atoll,
especially the foodstuffs, but it appears we will have to settle
for the external readings only.

cc: Dr. Western
    Dr. Seymour

NMP 4

OFFICIAL USE ONLY # 8

OFFICE ▶	BM						
FNAME ▶	GMDunning:em						
DATE ▶	6/13/57						

B18 (Rev. 9-53)   U. S. GOVERNMENT PRINTING OFFICE   16—62761-3

Letter from Hermann Lisco, MD, Cancer Research Institute, New England Deaconess Hospital, to George Darling, Director, Atomic Bomb Casualty Commission, April 29, 1966

CANCER RESEARCH INSTITUTE
NEW ENGLAND DEACONESS HOSPITAL
194 PILGRIM ROAD
BOSTON, MASSACHUSETTS 02215

April 29, 1966

Dr. George B. Darling
Director, Atomic Bomb Casualty
    Commission
U.S.Marine Corps Air Station
FPO San Francisco, California 96664

Dear Doctor Darling:

We have recently completed our studies on
chromosome abnormalities in the Marshallese people exposed to
fallout 12 years ago. We have examined all of the material that
was collected by Dr. R. A. Conard in the field two years ago.
A total of 51 persons has been examined and although the findings
are not spectacular, it appears that there is a difference between
the heavily exposed population and the controls.

In an effort to compare our findings to others, I find only
one reference to 7 Atomic Bomb Survivors in the open literature
(Doida et al, Rad.Res. 26, 69-83, 1965). I have not seen any other
reference to ABCC data either in the open literature or in the
Technical Report series. I am writing you to ask whether there
exist any other data on this population that could be used for
a comparison with ours. It is quite possible that I may have
overlooked something. I should be most grateful to you for any
information you can give me.

With kind regards,

Yours sincerely,

Hermann Lisco

Hermann Lisco, M. D.

Letter from Paul Seligman, U.S. Department of Energy, to Mayor James Matayoshi, Rongelap Atoll Local Government Council, April 29, 1999

Department of Energy
Germantown, MD 20874-1290

APR 29 1999

Mayor James Matayoshi
Rongelap Atoll Local Government Council
Republic of the Marshall Islands
P.O. Box 1766
Majuro, Marshall Islands 96960

Dear Mayor Matayoshi:

The Rongelap Atoll Local Government (RALGOV) has expressed interest in obtaining the names of Marshallese evaluated for anemia in the chromium-51 and tritiated water clinical tests conducted by Brookhaven National Laboratory (BNL) in the early 1960's. As a result of a recent consolidation of patient records from the Department of Energy's (DOE) Marshall Islands Medical Program into an electronic format, we now have an improved capability to search the database.

BNL's reports of the anemia tests are available on the website of DOE's Office of International Health Programs (EH-63) at http://tis.eh.doe.gov/ihp, and are among the 77 boxes of documents which DOE has provided to the Republic of the Marshall Islands Government.

BNL Report 908 (T-371) entitled, "Medical Survey of the People of Rongelap and Utirik Islands Nine and Ten Years After Exposure to Fallout Radiation (March 1963 and March 1964)," is document number 0403547 among the Marshall Islands Historical Documents on the EH-63 website. Table 20 of this Report, captioned "Total Blood and Red Cell Volume Data" (Enclosure 1), lists 20 of the 21 participants by "Subject No."; one participant is identified as "Jeton." Each "Subject No." corresponds to the medical record of a patient in DOE's Marshall Islands Medical Program. We are, therefore, able to identify all of the participants in the 1963 BNL anemia studies.

There is less complete information available for the 1961 and 1962 studies than for the 1963 studies, but we believe that 9 of the 13 Marshallese participants in the earlier studies can be identified.

BNL Report 780 (T-296) entitled, "Medical Survey of Rongelap People Eight Years After Exposure to Fallout (January 1963)," is document number 0403546 on the EH-63 website noted above. Table 15 of this report, captioned "Blood Studies (1961, 1962)" (Enclosure 2) lists 23 study participants, including 13 individuals identified as "Micronesian" and 10 as "Caucasian." Each participant is identified by initials (e.g., "T." or "Em.") and body weight. Furthermore, it appears from correspondence of the physician who conducted the studies that some 6 Marshallese participated in both groups of anemia studies. By correlating all of these data with age and weight information in DOE patient records, we have identified 9 Marshallese participants in the 1961 and 1962 blood volume studies utilizing chromium-51.

In compliance with considerations of personal privacy mandated by U.S. law, DOE can disclose this information directly to the individuals concerned. Therefore, we will notify each identified living DOE patient/participant in the 1961-1963 BNL anemia tests of this information. We will also provide information to the DOE/Pacific Health Research Institute Special Medical Care Program staff so that they can address questions patients might have during their visits to the DOE medical clinic at Kwajalein or Majuro.

Sincerely,

Paul J. Seligman, M.D., M.P.H.
Deputy Assistant Secretary
for Health Studies

2 Enclosures

cc w/enclosures:
Minister Phillip Muller, RMI
Ambassador Banny de Brum, RMI
Senator Johnsay Riklon, Rongelap
Howard Hills, Esq.
Ambassador Joan Plaisted

Glossary

177 Agreement: The section of the Compact of Free Association detailing U.S. responsibility for damages and injuries resulting from the U.S. nuclear weapons testing program in the Marshall Islands.

Advisory Committee on Biology and Medicine (ACBM): A committee of the Atomic Energy Commission that researched the impacts of radiation on human beings and the environment.

Advisory Committee on Human Radiation Experiments (ACHRE): A White House committee, established by executive order under President Bill Clinton, to investigate the extent to which U.S. government researchers used human beings in radiation experiments.

ak: A frigate bird.

alabs: Managers of an iroij's land who ensure that the land is used productively. *Alabs* protect the interests of the iroij and ensure that workers have rights to cultivate and live off the land.

alpha radiation: Created when two protons and two neutrons are emitted from the nucleus of an atom. Alpha particles have the same nucleus as the helium atom but lack the two electrons that make helium stable. Alpha particles travel up to 10,000 miles per second. Because they are so large in "subatomic" terms, alpha particles have been likened to large-caliber bullets. They tend to collide with molecules in the air and are easily slowed down. A thin sheet of paper or 2 inches of air can usually stop an alpha particle. When alpha-emitting elements are inhaled or ingested into the body, the high-energy particles they emit can rip into the cells of sensitive internal soft tissues, creating serious damage. Alpha particles are emitted by a wide array of heavy elements, including plutonium, a by-product of nuclear fission; radon, which seeps into the environment from uranium mining and milling processes; and radon gas, whose "daughter," or decay, elements are carried into the atmosphere from uranium-mining wastes. See also *Ionizing radiation*.

americium-241 (Am-241): Produced when plutonium atoms absorb neutrons in nuclear reactors and nuclear weapons detonations. It has a half-life of 432.7 years. All isotopes of americium are radioactive. As it decays, it releases alpha and gamma radiation and changes into eptunium-237, which is also radioactive. The americium-241 decay chain ends with bismuth-209, a stable (nonradioactive) element. People may be directly exposed to gamma radiation from americium-241 by walking on contaminated land. They may also be exposed to both alpha and gamma radiation by breathing in Am-241-contaminated dust, or drinking contaminated water. People who live or work near a contaminated site, such as a former weapons testing or production facility, may ingest americium-241 with food and water, or may inhale it as part of resuspended dust. In the human body, americium-241 tends to concentrate in the bone, liver, and muscle. When inhaled, some Am-241 remains in the lungs, depending upon the particle size and the chemical form of the americium compound. Chemical forms that dissolve easily may pass into the bloodstream from the lungs. Chemical forms that dissolve less easily tend to remain in the lungs or are coughed up through the lung's natural defense system and swallowed. From the stomach, swallowed americium may dissolve and pass into the bloodstream. Undissolved material passes from the body through the feces. Americium-241 can stay in the body for decades, exposing the surrounding tissues to both alpha and gamma radiation and increasing the risk of cancer. It also poses a cancer risk to all organs of the body from direct external exposure to its gamma radiation.

arrowroot (*Maranta arundinacea*): A starchy staple of the Marshallese diet that is easy to preserve and digest.

atomic bomb: An explosive device in which a large amount of energy is released through the nuclear fission of uranium or plutonium. The first atomic bomb test, known as the Trinity shot, took place in the desert north of Alamogordo, New Mexico, on July 16, 1945. Several weeks later, an atomic bomb was used for the first time as an instrument of war, detonating over the Japanese cities of Hiroshima (August 6) and Nagasaki (August 9). The next detonation occurred in the Marshall Islands during the 1946 Operation Crossroads, in which atmospheric (Able) and subsurface (Baker) tests deployed Nagasaki-type bombs.

Atomic Bomb Casualty Commission (ABCC): The commission established by President Harry Truman in 1946 to study the effects of radiation on human beings and the environment after the bombing of Hiroshima and Nagasaki.

Atomic Energy Commission (AEC): The predecessor to the U.S. Department of Energy, established by the U.S. Congress in 1946. The AEC became the Department of Energy in 1974.

atoll: A series of coral islands with a lagoon in the middle.

background radiation: Ionizing radiation from both natural and man-made sources. Background radiation includes cosmic radiation; radiation emitted by naturally occurring radionuclides in air, water, soil, and rock; radiation emitted by natural radionuclides deposited in tissues of organs; and ionizing radiation from man-made sources, such as nuclear weapons fallout, nuclear power operations and accidents, and diagnostic x-rays and consumer products. Exposures to background radiation vary depending on the geographic area, diet, and other factors, such as the composition of materials used in the construction of homes.

Becquerel (Bq): The International System (SI) unit for activity of radioactive material. One Bq of radioactive material is that amount in which one atom is transformed or undergoes one disintegration every second. Whole-body counting and plutonium bioassay measurements are usually reported in activity units of kBq (kilo-Becquerel) (1,000 x 1 Bq) and μBq (micro-Becquerel) (1 x 10-6 x 1 Bq), respectively. See also *Units of radioactivity*.

beta radiation: Composed of streams of electrons that often travel at close to the speed of light. In some cases, beta particles are emitted from a nucleus when a neutron breaks down into a proton and electron. The proton stays in the atom's core while the electron shoots out. Because they move faster than alpha particles, and weigh much less, beta particles are far more penetrating than alpha particles. Sheets of metal and heavy clothing are required to stop them. Beta emissions to the skin can lead to skin cancer. Like elements that emit alpha particles, beta emitters can be very dangerous when inhaled or ingested into the body. Beta radiation can be emitted from many substances released by nuclear bombs and power plants, including strontium-90 and tritium. See also *Ionizing radiation*.

bioaccumulation: The increase in concentration of a substance in an organism over time.

biodistribution: The pattern and process of a chemical substance's distribution through the body.

biopsy: The removal and/or examination of tissues, cells, or fluids from a living body for the purposes of diagnosis or experimental tests.

body burden: The amount of a radioactive material present in a body over a long period. It is calculated by considering the amount of material initially present and the reduction in that amount due to elimination and radioactive decay. It is commonly used in reference to radionuclides having long biological half-lives. A body burden that subjects the body's most sensitive organs to the highest dose of a particular radionuclide that regulators allow is known as a maximum permissible body burden (MPBB).

bone marrow: The soft tissue contained within the internal cavities of the bones. Bone marrow is a site of blood-cell formation, especially in young animals and humans.

bone-marrow infusion: The injection of bone marrow (an essential tissue producing red and white blood cells and platelets) into the body; used primarily to replace bone marrow destroyed by disease or in the course of radiation and other therapies for certain types of cancer.

Brookhaven National Laboratory (BNL): The U.S. weapons laboratory in Brookhaven, New York, contracted by the U.S. government to conduct medical and scientific research on the consequences of radiation exposure in the Marshall Islands. Brookhaven played a key role in establishing and maintaining Project 4.1.

bwebwenato: Storytelling, talking together.

bwij: Literally meaning "land" but used to refer to lineage (because land is the basis of lineage in the Marshall Islands).

calibration: The process of adjusting or determining the response or reading of an instrument to a standard.

cancer: A general term for more than one hundred diseases that involve uncontrolled, abnormal growth of cells that can invade and destroy healthy tissues.

carcinogen: A material that can initiate or promote the development of cancer. Well-known carcinogens include saccharine, nitrosamines found in cured meat, certain pesticides, and ionizing radiation.

cesium-137: Nonradioactive cesium occurs naturally in various minerals. Radioactive cesium-137 is produced when uranium and plutonium absorb neutrons and undergo fission. Cesium-137 has a half-life of 30 years. It decays by emission of a beta particle and gamma rays to barium-137m, a short-lived decay product, which in turn decays to a nonradioactive form of barium. Because of the chemical nature of cesium, it moves easily through the environment and bioaccumulates in plants and animals. People may ingest cesium-137 with food and water, or may inhale it as dust. If cesium-137 enters the body, it is distributed fairly uniformly throughout the body's soft tissues, resulting in exposure of those tissues. Slightly higher concentrations of the metal are found in muscle, while slightly lower concentrations are found in bone and fat. Compared to some other radionuclides, cesium-137 remains in the body for a relatively short time. It is eliminated through the urine. Exposure to cesium-137 may also be external (that is, exposure to its gamma radiation from outside the body). As with all radionuclides, exposure to radiation from cesium-137 results in increased risk of cancer. If exposures are very high, serious burns and even death can result. The magnitude of the health risk depends on exposure conditions. These include such factors as strength of the source, length of exposure, distance from the source, and whether there was shielding between a person and the source (such as metal plating).

chromium-51 (Cr-51): an ion of the element chromium, which is used as a radioactive tracer because it binds to red blood cells. It is a gamma emitter with a 27.7-day half-life. By measuring the amount of radioactivity produced by Cr-51, red blood cell mass and survival can be measured. Exposure can occur via ingestion, inhalation, and skin absorption. Once absorbed, Cr-51 tends to concentrate in the liver, bones, large intestines, and kidneys.

chromosome: Each chromosome is composed of deoxyribonucleic acid (DNA) and specialized protein molecules, which convey genetic information. They are located in the nucleus of both plant and animal cells. In humans there are forty-six chromosomes.

chromosome aberration: Any deviation from the normal number or morphology of chromosomes.

changed circumstances: A provision of the Compact of Free Association allowing the RMI government to petition the U.S. Congress for additional assistance resulting from damages and injuries from the testing program not known or understood during bilateral negotiations of the compact.

chelating therapy: One of the human radiation experiments in the Marshall Islands confirmed by ACHRE. The therapy binds chelating agents (such as EDTA) to heavy metals to assist with the removal of these metals from the body.

ciguatera: Fish poisoning created when reef fish eat smaller fish and bioaccumulate the neurotoxins produced by the consumption of toxic algae.

coconut crab (*Birgus latro*): The largest terrestrial arthropod in the world. It is a hermit crab and is known for its ability to crack coconuts with its strong pincers so it can eat the contents. The coconut crab bioaccumulates radiation and is a favorite food of the Marshallese.

Compact of Free Association: U.S. Public Law 99-239 (amended in 2003 under U.S. Public Law 180-188) defines the terms of the bilateral relationship between the United States and the RMI. The compact provides economic assistance to the RMI in exchange for critical U.S. defense rights. For example, the United States retains the right to deny third-country military vessels into the Marshall Islands, and Marshallese men and women serve in every branch of the U.S. armed forces.

Congress of Micronesia: Established in 1966 by President Johnson to help citizens of the Trust Territory of the Pacific Islands achieve greater self-determination. The congress paved the way for the termination of trusteeship twenty years later.

copra: Dried coconut meat and the most important cash-generating commodity for Marshallese living on rural islands. Making copra is labor intensive. It requires gathering and opening coconuts by hand, removing and drying the meat, and bagging it for sale to middlemen, who arrive in boats to bring copra to the capital.

C-rations (combat rations): Packaged meals used by the U.S. armed forces.

Curie (Ci): One curie is the quantity of a radioactive material that will have 37 billion disintegrations per second (1 Ci = 37 billion Bq). See *Units of radioactivity*.

customary laws: Traditional practices that reflect the priorities of a culture, such as land-distribution practices in the RMI. In the Marshall Islands, customary law is codified in the national constitution.

danger area: The area in the RMI determined to be in danger of high radiation exposure during test events. During the Bravo event, on March 1, 1954, the U.S. government excluded the inhabited atolls of Rongelap and Ailinginae from the danger area at the last minute, despite wind patterns blowing from the test site to those atolls.

Defense Nuclear Agency (DNA): A U.S. agency established to support nuclear activities during the Cold War. The DNA, which no longer exists, constructed a nuclear waste storage facility on Runit Island in Enewetak Atoll. No U.S. agency currently has responsibility for monitoring this facility.

Department of Energy (DOE): The successor to the AEC. The agency has responsibility for contracting medical and environmental programs in the RMI that address that damages and injuries resulting from the testing program that the U.S. government currently accepts responsibility for.

deterministic effect: Radiation or other agents can create a deterministic effect, or increased severity. Kidney damage is one example of a deterministic effect.

diagnostic procedure: A method used to identify a disease in a living person.

dose: In radiology, a measure of energy absorbed in the body from ionizing radiation; measured in the form of rads. See *rad*.

dose equivalent: The absorbed radiation dose adjusted to consider the biological harmfulness of different kinds of radiation; measured in sieverts (Sv).

dose rate: The ionizing radiation dose delivered per unit of time.

dose reconstruction: The process of using information about an individual's past exposures to ionizing radiation, as well as general knowledge about the behavior of radioactive materials in the human body and in the environment, to estimate the dose of radiation that someone has received.

dosimeter: An instrument that measures the dose of ionizing radiation. A biological dosimeter is a biological or biochemical indicator of the effects of exposure, such as a change in blood chemistry or blood count. A highly accurate biological dosimeter has yet to be found.

drekeinin: A mallet made from clamshells, used to bang dried pandanus to soften the long leaves and make them suitable to weave into sleeping mats. Mallets are passed from generation to generation matrilineally.

EDTA (ethylenediamminetetraacetate): See *chelating therapy*.

emok: A giant clamshell used for water storage.

endocrine glands: The endocrine glands manufacture one or more hormones and secrete them directly into the bloodstream. Endocrine glands include the pituitary gland, thyroid, parathyroid, adrenal glands, ovary, testis, placenta, and part of the pancreas.

exclusive economic zone (EEZ): A law of the sea term denoting a marine area in which a state possesses rights to explore and cultivate marine resources.

exposed (versus nonexposed): Section 177 of the Compact of Free Association defines exposed people as the people present on Rongelap, Ailinginae, and Utrik atolls on March 1, 1954. The "exposed" population is legally eligible to participate in U.S. health care provided by DOE. The term is controversial because it ignores evidence suggesting that the entire nation was exposed to harmful levels of radioactive fallout.

fallout: Radioactive debris that falls to earth after a nuclear explosion.

film badges: Badges developed by the U.S. armed forces to record individual levels of radiation exposure. Badges were not used regularly by U.S. military personnel in the Marshall Islands and often did not accurately record acute exposure levels. Many U.S. servicemen suffered severe health consequences from radiation exposure in the RMI. No film badges were ever distributed to Marshallese citizens.

fish poisoning: See *ciguatera*.

fission: The division of an atomic nucleus into parts of comparable mass. Generally speaking, fission may occur only in heavier nuclei, such as isotopes of uranium and plutonium. Atomic bombs derive energy from the fission of uranium or plutonium.

fission product: An atom or nucleus that results from the fission of a larger nucleus.

fusion: The combining of two light atomic nuclei to form a single heavier nucleus, releasing energy. Hydrogen bombs derive a large portion of their energy from the fusion of hydrogen isotopes.

gamma radiation: A form of electromagnetic or wave energy similar in some respects to x-rays, radio waves, and light. Like x-rays, gamma radiation is highly energetic and can penetrate matter much more easily than alpha or beta particles. Gamma rays are usually emitted from the nucleus when it undergoes transformations. An inch of lead or iron, 8 inches of heavy concrete, or 3 feet of sod may be required to stop most of the gamma rays from an intense source. See also *ionizing radiation*.

gene: The functional unit of heredity that occupies a specific place on a chromosome.

genetic effects: Changes in a person's germ cells (sperm or ova) that are transmissible to future generations. Such changes result from mutations in genes within the germ cells.

gray (Gy): Measures a quantity of radiation energy called the absorbed dose, which is the amount of energy actually absorbed in any given material (1 Gy = 100 rads). See *units of radioactivity*.

half-life: The average time required for one-half of the amount of radioactivity of a radionuclide to undergo radioactive decay. For material with a half-life of one week, half of the original amount of activity will remain after one week; half of that (one-quarter of the original amount) will remain after two weeks, and so on.

health physics: A branch of physics specializing in accurate measurement of agents, such as ionizing radiation, that effect human health.

hematocrit: The percentage of the volume of a blood sample occupied by cells, as determined by a centrifuge or device that separates the cells and other particulate elements of the blood from the plasma. The remaining fraction of the blood sample is called plasmocrit (blood plasma volume).

hemoglobin: An iron-containing respiratory pigment contained within red blood cells; it gives the cells their red color. Hemoglobin, which has the unique property of combining reversibly with oxygen, picks up oxygen in the lungs and transports it to the rest of the body.

hormone: A substance produced in one part of the body; it passes into the bloodstream and is carried to distant organs or tissues, where it acts to modify their structure or function.

hydatidiform mole: A rare mass or growth that forms inside the uterus at the beginning of a pregnancy. A hydatidiform mole results from overproduction of the tissue that is supposed to develop into the placenta. The placenta normally feeds a fetus during pregnancy. In this condition, the tissues develop into an abnormal growth, called a mass. Often, there is no fetus at all. In 10 to 15 percent of cases, hydatidiform moles may develop into invasive moles. These moles may grow far into the uterine wall and cause bleeding or other complications. In a few cases, a hydatidiform mole may develop into a choriocarcinoma, a fast-growing, spreading form of cancer. A hydatiform mole is referred to as a grape pregnancy by the Marshallese.

hydrogen bomb (H-bomb): An explosive weapon, also known as a thermonuclear bomb, that uses nuclear fusion to release energy stored in the nuclei of hydrogen isotopes. The high temperatures essential to fusion are attained by detonating an atomic bomb placed at the H-bomb's structural center. The United States tested its largest hydrogen bomb (Bravo test) in 1954 at the Pacific Test Site.

immune system: The immune system provides the body with a defense against infection, afforded by the presence of circulating antibodies and white blood cells. Antibodies are manufactured specifically to deal with antigens associated with different diseases. White blood cells attack and destroy foreign particles in the blood and tissues, including antigen-antibody complexes. Exposure to ionizing radiation adversely affects the immune system.

influenza virus: An acute infectious respiratory disease in which the inhaled virus attacks the respiratory epithelial cells of susceptible persons and produces an inflammation of the mucous membrane. Influenza virus is of the genus *Orthomyxoviridae*, which comprises the type-A and -B influenza viruses. Each type of virus has a stable nucleoprotein group antigen common to all strains of the type but distinct from that of the other type; each also has a mosaic of surface antigens (hemagglutinin and neuraminidase) that characterize the strains. Strain notations indicate (1) type; (2) geographic origin; (3) year of isolation; and (4), in the case of type-A strains, the characterizing subtypes of hemagglutinin and neuraminidase antigens (for example, A/Hong Kong/1/68 [H3N2]).

internal emitter: A radioisotope incorporated into a tissue in the body that decays in place and continuously exposes that tissue to ionizing radiation.

iodine isotopes: Iodine is a nonmetallic, purplish-black crystalline solid. It has the unusual property of sublimation, which means that it can go directly from solid to gas without first becoming liquid. Iodine reacts easily with other chemicals, and isotopes of iodine are found as compounds rather than as a pure elemental nuclide. Radioactive iodines are produced by the fission of uranium atoms during operation of nuclear reactors and by plutonium (or uranium) in the detonation of nuclear weapons. They have the same physical properties as stable iodine, easily bonding with chemical compounds, and they emit beta particles as they decay. Iodine-129 has a half-life of 15.7 million years; iodine-131 has a half-life of about 8 days. Because of its short half-life and useful beta emission, iodine-131 is used extensively in nuclear medicine. Its tendency to collect in the thyroid gland makes iodine especially useful for diagnosing and treating thyroid problems. Iodine-123 is widely used in medical imaging, and I-124 is useful in immunotherapy. Iodine's chemical properties make it easy to attach to molecules for imaging studies. It is useful in tracking the metabolism of drugs or compounds, and for viewing structural defects in various organs, such as the heart. A less common isotope, iodine-125, is sometimes used to treat cancerous tissue. Iodine-129 has little practical use but may be used to check some radioactivity counters in diagnostic testing laboratories.

Iodine-129 and iodine-131 are gaseous fission products that form within reactor fuel rods as they fission, and in the detonation of nuclear weapons. Under the right conditions, radioactive iodine can disperse rapidly in air and water. However, it combines easily with organic materials in soil. This process is known as organic fixation, and it slows iodine's movement in the environment. Some soil minerals also attach to or adsorb iodine, which also slows its movement. The long half-life of iodine-129, 15.7 million years, means that it remains in the environment. However, iodine-131's short half-life of 8 days means that it will decay away completely in the environment in a matter of months. Both decay with the emission of a beta particle, accompanied by weak gamma radiation. Radioactive iodine can be inhaled as a gas or ingested in food or water. It dissolves in water, so it moves easily from the atmosphere into humans and other living organisms. For example, I-129 and I-131 can settle on grass, where cows can eat it and pass it to humans through milk. It may settle on leafy vegetables and be ingested by humans. Iodine isotopes also concentrate in marine and freshwater fish, which people may then eat. Also, doctors may give thyroid patients radioactive iodine, usually iodine-131, to treat or help diagnose certain thyroid problems. The tendency of iodine to collect in the thyroid makes it very useful for highlighting parts of its structure in diagnostic images. When I-129 or I-131 is ingested, some of it concentrates in the thyroid gland. The rest passes from the body in urine. Airborne I-129 and I-131 can be inhaled. In the lung, radioactive iodine is absorbed; it passes into the bloodstream and collects in the thyroid. Any remaining iodine passes from the body with urine. In the body, iodine has a biological half-life of about 100 days for the body as a whole. It has different biological half-lives for various organs: thyroid, 100 days; bone, 14 days; kidney, spleen, and reproductive organs, 7 days. Long-term (chronic) exposure to radioactive iodine can cause nodules, or cancer of the thyroid. However, once thyroid cancer occurs, treatment with high doses of I-131 may be used to treat it. Lower doses of I-131 may be used to treat overactive thyroids by reducing activity of the thyroid gland and lowering hormone production in the gland. The thyroid cannot tell the difference between radioactive and nonradioactive iodine. It will take up radioactive iodine in whatever proportion it is available in the environment. If large amounts of radioactive iodine are released during a nuclear event, government agencies may distribute large doses of stable iodine to keep people's thyroid glands from absorbing too much radioactive iodine. Raising the concentration of stable iodine in the blood increases the likelihood that the thyroid will absorb it instead of radioactive iodine. Large doses of stable iodine can be a health hazard and should not be taken except in an emergency. However, iodized table salt is an important means of acquiring essential nonradioactive iodine to maintain health.

ionization: The process by which a neutral atom or molecule loses or gains electrons, thereby acquiring a net electrical charge. When charged, it is known as an ion.

ionizing radiation: Any of the various forms of radiant energy that cause ionization when they interact with matter. The most common types are alpha radiation, made up of helium nuclei; beta radiation, made up of electrons; and gamma and x radiation, consisting of high-energy particles of light (photons).

iroij: Chiefs who customarily assume ownership of the land on behalf of their generation. Iroij are responsible for ensuring the well-being of all people who live and work on their land. Alaps manage the land on behalf of the iroij.

Iroij Rilik: God of fish.

iron: Iron-55 (Fe-55) has a 2.68-year half-life. Iron-59 (Fe-59) has a 44.5-day half-life.

Exposure can occur via ingestion, inhalation, puncture, and wound and skin absorption. Once absorbed, iron tends to concentrate in the spleen and blood.

irradiation: Exposure to radiation of any kind, especially ionizing radiation.

isotope: A species of nucleus with a fixed number of protons and neutrons. The term *isotope* is usually used to distinguish nuclear species of the same chemical element (that is, those having the same number of protons but different numbers of neutrons), such as iodine-127 and iodine-131. Atoms with the same number of protons but different numbers of neutrons are called isotopes of that element. Different isotopes are identified by appending the total number of nucleons (the total number of protons and neutrons in the nucleus of an atom) to the name of the element—for example, cesium-137. Isotopes are usually written in an abbreviated form using the chemical symbol of the element. Two examples are Cs-137 for cesium-137 and Pu-239 for plutonium-239.

janwin: Preserved breadfruit that allowed the Marshallese to have a more diverse diet during times when major types of breadfruit were out of season.

Jebro: God of breadfruit.

jekaka: Dried pandanus; suitable for grating, and easy to store.

jibun: Miscarriage.

Joint Task Force 7 (JTF-7): The U.S. Department of Navy team responsible for carrying out the testing of atomic and thermonuclear devices in the Marshall Islands.

jolet: A social group determined by its collective inheritance of common land.

jujukop: Barracuda fish.

kajor: A synonym for *ri-jerbal,* literally meaning "strength"; refers to workers who cultivate the land and keep it productive for the iroij.

kalo: Brown booby (bird).

kano: Fern.

kear: Tern.

kejinbwij: The collecting and distribution of foods by an iroij.

kemem: A family and community birthday celebration for a one-year old.

kinbit: Rules and regulations for collecting food, such as coconut crabs.

kitde **land:** Land that an iroij or alap gives to his wife. The land passes down through the family of the wife upon her death.

K-rations: Three meals bundled into a single package; designed to sustain a U.S. Army soldier for one day during World War II.

Kwajalein Atoll: The world's largest atoll; home to the U.S. Army's Ronald Reagan Ballistic Missile Defense Test Site on Kwajalein Island, as well as a portion of the Rongelap community residing on Mejatto Island.

latency period: The time between when an exposure to radiation occurs and when its effects are detectable as an injury or illness.

lagoon: In the RMI, the lagoon is the enclosed water in the center of a circular string of small coral islands. The water is calmer and shallower than ocean. In rural areas, most homes are built on the lagoon side of the islands.

lamoren: Lineage land extending back for generations. The term denotes a sense of eternal heritage and the essence of what it means to be Marshallese.

laroij: A female iroij, or chief.

Lawi Jemo: Kanal tree god.

leukemia: Any group of malignant diseases in which the bone marrow and other blood-forming organs produce increased numbers of leukocytes (white blood cells).

lineage: See *bwij.*

lojepjep: grouper fish.

LST: Landing ship, tank; a ship that can be beached for the easy unloading of cargo and crew. LSTs helped with the relocation of populations in the Marshall Islands either as precautionary measures in advance of weapons tests or following the exposure of populations to radiation from weapon events.

Majuro: The name of the RMI's capital, Majuro Island, as well as Majuro Atoll, where the island is located.

metabolism: The manner in which a substance is acted upon (taken up, converted to other substances, and excreted) by various organs of the body.

millirem: One-thousandth of a rem. See *units of radiation.*

millisievert: A standard international unit for measuring dose to humans (1 millisievert = 100 millirems). See *units of radiation.*

mo: Forbidden.

mon-tutu: Shower house.

morjinkôt **land:** Land, such as Rongerik Atoll, given by an iroij to a warrior for heroics in battle.

mweo: A type of long line fishing.

mwilmwil: Mackerel.

National Radiation Commission (NRC): The U.S. National Radiation Commission has responsibility for exploring civilian applications of radioactive materials in ways that do not compromise the well-being of people or the environment.

natural background radiation: Ionizing radiation that occurs naturally. Its principle sources are cosmic rays from outer space, radionuclides in the human body, and radon gas (a decay product of natural uranium in the earth's crust).

National Cancer Institute (NCI): The component of the U.S. National Institutes of Health (NIH) tasked with reducing the burden of cancer through research and training.

Nitijela: The Marshallese parliament.

ni: A young, drinking coconut.

Nuclear Claims Tribunal (NCT): The judiciary forum established by the 177 Agreement of the Compact of Free Association to consider personal injury and property-damage claims resulting from the U.S. nuclear weapons testing program in the Marshall Islands. The tribunal was meant to serve as an alternative to the U.S. courts.

ok: Net for fishing.

Operation Castle: A series of six high-yield tests conducted by JTF-7 on Bikini Atoll in 1954, including the infamous Bravo event on March 1, 1954.

Pandanus tectorius: A wide-branched tree that grows to heights of about 8 meters (25 feet). One of a few edible species in the Marshall Islands, pandanus is a very important plant that has been cultivated for thousands of years. It is used for food, medicine, thatch for traditional houses, and fiber for mats, hats, and baskets. In the past, canoe sails were plaited from the leaves.

pejwak: Brown noddy (bird; *Anous stolidus*).

permissible dose: In the judgment of a regulatory or advisory body, such as the National Committee on Radiation Protection, the amount of radiation that may be received by an individual within a specified period.

PHRI (Pacific Health Research Institute): Located in Hawaii, the first non-U.S.- government health care delivery group to assume responsibility for the DOE medical program after Brookhaven National Laboratory. The human research aspect of the medical program terminated when PHRI assumed responsibility for the care of the "exposed" Marshallese.

Plutonium: Created from uranium in nuclear reactors and used in nuclear weapons. When uranium-238 absorbs a neutron, it becomes uranium-239 before decaying to plutonium-239. Various isotopes of uranium and different combinations of neutron absorptions and radioactive decay create at least fifteen types of plutonium isotopes, all of which are radioactive. The most common ones are Pu-238, Pu-239, and Pu-240. Pu-238 has a half-life of 87.7 years. As plutonium decays, it releases radiation and forms other radioactive isotopes. For example, Pu-238 emits an alpha particle and becomes uranium-234; Pu-239 emits an alpha particle and becomes uranium-235. This process happens slowly, since the half-lives of plutonium isotopes tend to be relatively long: plutonium-239 has a half-life of 24,100 years; Pu-240 has a half-life 6,560 years. Plutonium-239 is used to make nuclear weapons. The plutonium in the bomb undergoes fission in an arrangement that assures enormous energy generation and destructive potential. Plutonium was dispersed worldwide from atmospheric testing of nuclear weapons conducted during the 1950s and 1960s. The fallout from these tests left very low concentrations of plutonium in soils around the world and high concentrations at nuclear weapons production and testing facilities. People may inhale plutonium as a contaminant in dust. It can also be ingested with food or water. The stomach does not absorb plutonium very well, and most plutonium swallowed with food or water passes from the body through the feces. Plutonium is most readily absorbed through mucus membranes. When inhaled, plutonium can remain in the lungs, depending upon its particle size and how well the particular chemical form dissolves. The chemical forms that dissolve less easily may lodge in the lungs or move out with phlegm and then either be swallowed or spit out. But the lungs may absorb chemical forms that dissolve more easily and pass them into the bloodstream. Once in the bloodstream, plutonium moves throughout the body and into the bones, liver, and other organs. Plutonium that reaches body organs generally stays in the body for decades and continues to expose the surrounding tissue to radiation. External exposure to plutonium poses very little health risk, since plutonium isotopes emit alpha radiation and almost no beta or gamma radiation. In contrast, internal exposure to plutonium is an extremely serious health hazard. It generally stays in the body for decades, exposing organs and tissues to radiation and increasing the risk of cancer. In addition to its radioactive nature, plutonium is also a toxic metal and may cause damage to the kidneys.

plankton: Any drifting organism that inhabits the water column of oceans, seas, and bodies of freshwater.

principal investigator: The scientist or scholar with primary responsibility for the design and conduct of a research project.

Project 4.1: A top-secret medical research program to study the effects of radiation exposure on human beings exposed to radiation from the Bravo event on March 1, 1954.

Project SHAD: From 1963 through the early 1970s, the U.S. Department of Defense conducted tests to determine the effectiveness of shipboard detection of and protective measures against both chemical and biological warfare agents, and less toxic simulations of these agents. The tests were conducted under the broad heading of Shipboard Hazard and Defense (SHAD), part of a larger activity called Project 112 that included similar land-based tests.

protocol: The formal design or plan of an experiment or research activity; specifically, the plan submitted to an institutional review board for review and to a government agency for research support. Protocols include a description of the research design or methodology to be employed, the eligibility requirements for prospective subjects and controls, the treatment regimen(s), and the methods of analysis to be performed on the collected data.

psychosocial stigmatization: Individual and communitywide experience with and feelings of alienation, social unacceptability, shame, or disgrace.

rad: Short for "radiation absorbed dose." A measurement of the amount of radiation absorbed by tissues; 1 rad is the amount of radiation that will deposit 1/100 joule per kilogram (100 rads = 1 Gy). See *units of radiation*.

radiation: The emission of waves transmitting energy through space or a material medium, such as water. Light, radio waves, and x-rays are all forms of radiation. When a radioactive particle or ray strikes a cell, one of at least four things can happen: It may pass through the cell without doing any damage; it may damage the cell but in such a way that the cell can recover and repair itself before it divides; it may kill the cell; or it may damage the cell in such a way that the damage is repeated when the cell divides.

radiation dose (mrem): A generic term to describe the amount of radiation a person receives. Dose is measured in units of thousands of a roentegen equivalent in man (rem). The conventional unit used by federal and state agencies in the United States is the millirem (mrem). Dose is a general term used to assist in the management of exposure to radiation. The common International System (SI) unit for dose is the millisievert (mSv); (1 mSv = 100 mrem).

radiation sickness: Acute physical illness caused by exposure to doses of ionizing radiation large enough to cause toxic reactions. This sickness can include symptoms such as nausea, diarrhea, headache, lethargy, and fever.

radioactive decay: The process by which the nucleus of a radioactive isotope decomposes and releases radioactivity. For example, carbon-14 (a radioisotope of carbon) decays by losing a beta particle, thereby becoming nitrogen-14, which is unstable.

radioactive iodine: See *iodine isotopes.*

radioactivity: The decay of unstable nuclei through the emission of ionizing radiation. The resulting nucleus may itself be unstable and undergo radioactive decay. The process stops only when the decay product is stable.

radiogenic: A term used to identify conditions observed to be caused by exposure to ionizing radiation, such as certain kinds of cancer.

radioisotope: A radioactive isotope of an element. Radioisotopes are used in medical research as tracers. See also *isotope, nuclide,* and *radionuclide.*

radiological weapons: Weapons that use radioactive materials to cause radiation injury.

radionuclide: A radioactive nuclide, often used to distinguish radioisotopes of different chemical elements, such as iodine-131 and uranium-239. In the human body, radionuclides that are in soluble form and chemically analogous to essential nutrient elements will typically follow pathways in a fashion similar to their nutrient analogues. For example, Sr-89, Sr-90, Ba-140, Ra-226, and Ca-45 behave like calcium and are bone-seeking elements; Cs-137, Rb-86, and K-40 follow the general movement of potassium and will be found throughout the body; I-129 and I-131 behave like stable iodine and accumulate in the thyroid; tritium resembles hydrogen and, as tritiated water, will be distributed throughout the body. Elements that demonstrate unique behavior include Ce-144, Ru-106, Zr-95, Kr-85, and Pu-239.

radiopharmaceuticals: Drugs (compounds or materials) that may be labeled or tagged with a radioisotope. In many cases, these materials function much like materials found in the body and do not produce special pharmacological effects. The principal risk associated with these materials is the consequent exposure of the body or certain tissues to radiation.

radio-resistance: The degree of resistance of organisms or tissues to the harmful effects of ionizing radiation.

radiosensitivity: The degree of sensitivity of organisms or tissues to the harmful effects of ionizing radiation.

RALGOV: The abbreviation for Rongelap local government, which serves the interests of Rongelap, Rongerik, and Ailinginae atolls.

Ralik: The RMI consists of two north-to-south chains of atolls. Ralik, meaning "sunset," is the chain to the west.

Ratak: The RMI consists of two north-to-south chains of atolls. Ratak, meaning "sunrise," is the chain to the east.

rem: Short for "roentgen equivalent man"; a unit used to derive a quantity called the equivalent dose. This unit of measurement reflects the fact that not all radiation has the same effect on living human tissue, even for the same amount of absorbed dose (1 rem = .01 Sv = 1,000 mrem [millirem]). See also *units of radiation*.

rep: See *units of radiation*.

RMI (Republic of the Marshall Islands): The RMI became an independent nation in 1986, when the Compact of Free Association came into effect. The nation was no longer part of the United Nations Trust Territory of the Pacific Islands (TTPI), administered by the United States.

ri-jerbal: See *kajor*.

risk: The probability of harm from the presence of radionuclides or hazardous materials, taking into account (1) the probability of occurrences or events that could lead to an exposure; (2) probability that individuals or populations would be exposed to radioactive or hazardous materials and the magnitude of such exposures; and (3) the probability that an exposure would produce a response.

roentgen (R): Measures the ability of photons (gamma rays and x-rays) to make ions in the air (not in tissue or other materials). See also *units of radiation.*

roro: A chant used to disseminate information orally or to document history.

seamounts: Large submarine volcanic mountains rising at least 1,000 meters (3,300 feet) above the surrounding deep-sea floor. Preliminary oceanographic studies have shown that seamounts are biologically rich areas supporting a distinct benthic (bottom-dwelling) community of animals, many of which are unique and do not occur elsewhere on earth. In addition to supporting diverse marine life, seamounts attract pelagic species—schools of large fish (especially tuna)—that visit to feed and spawn.

sievert (Sv): A measurement of equivalent dose in humans (1 Sv = 100 rem).

staphylococcal enterotoxin B (PG): Produced by the bacteria *Staphylococcus aureus;* used by the United States as an anthrax-simulating agent in biological warfare experiments. Exposure can cause illness if inhaled in low doses. The toxin interacts with the individual's immune system to produce a variety of effects. Symptoms of inhaled SEB appear three to twelve hours after exposure and can include sudden onset of a high fever (103–106° F), chills, headache, muscle aches, a dry cough, and inflammation of the lining of the eyelids. There may also be difficulty breathing, chest pain, fluid in the lungs, or a fever for two to five days. The cough may persist up to four weeks. Exposure will result in 80 percent incapacitation of the population. Above symptoms can also occur via absorption through the skin; such exposure may result in inflammation of the lungs, pneumonia, and death in the very young or old. If the toxin is swallowed, there may be nausea, vomiting, and diarrhea, with no symptoms involving the lungs.

strontium-90: A by-product of the fission of uranium and plutonium in nuclear reactors and nuclear weapons. As strontium-90 decays, it releases radiation and forms yttrium-90 (Y-90), which in turn decays to stable zirconium. The half-life of Sr-90 is 29.1 years, and that of yttrium-90 is sixty-four hours. Sr-90 emits moderate-energy beta particles, and Y-90 emits very strong (energetic) beta particles. Strontium-90 can form many chemical compounds, including halides, oxides, and sulfides, and moves easily through the environment and into the human food chain. People may also inhale trace amounts of strontium-90 as a contaminant in dust. But swallowing Sr-90 with food or water is the primary pathway of intake. Strontium-90 is chemically similar to calcium and tends to deposit in teeth, bone, and blood-forming tissue (bone marrow). Thus strontium-90 is referred to as a bone seeker. Internal exposure to Sr-90 is linked to bone cancer, cancer of the soft tissue near the bone, and leukemia.

Superfund: The name of a U.S. environmental program established to address abandoned hazardous waste sites. It is also the name of the fund established by the Comprehensive Environmental Response, Compensation and Liability Act of 1980, as amended, which allows the Environmental Protection Agency to clean up hazardous waste sites, compel responsible parties to perform cleanups, and fine responsible parties to reimburse the government costs of cleanups.

taban: Medicinal areas that are off-limits except to *ri-bubu*, or healers.

taboos: Rules that restrict behavior. Taboos are defined and imposed by internal actors in ways that establish and regulate social, political, and economic relationships. While *taboo* is not a Marshallese word, the Marshallese use the term when speaking in English about those "rules that restrict behavior." In the Marshall Islands, rules are known and followed, and there are specific Marshallese terms for specific sets of rules. But adherence to these rules is not enforced by a strict code of sanctions, as has been described in some other Pacific Islands cultures.

TERPACIS: A military term used to denote the Trust Territory of the Pacific Islands.

teratogenic effects: Nonhereditary effects from an agent that are seen in the offspring of an individual who was exposed to the agent. The agent must be encountered during the gestation period.

tolerance dose: See *permissible dose.*

total-body irradiation (TBI): Exposure of the entire body to external radiation.

tracer: A distinguishable substance, usually radioactive, added to a nuclear weapon to develop a distinctive signature that allows tracking of weapons fallout locally, regionally, and worldwide. In the Marshall Islands, nuclear weapons fallout contained radioisotopes specific to the makeup and detonation of each bomb, as well as additional radioisotopes of added tracer elements. Reported tracers include sulfur, arsenic, yttrium, rhodium, indium, tantalum, tungsten, gold, thallium, polonium-210, thorium-230, thorium-232, uranium-233, uranium-238, americium-241, and curium-242.

tritium: A radioactive isotope of hydrogen. Tritium contains one proton and two neutrons in its nucleus. Because it is chemically identical to the natural hydrogen atoms present in water, tritium can easily be taken into the body by ingestion. It decays by beta emission and has a radioactive half-life of about 12.3 years. As it undergoes radioactive decay, tritium emits a very low-energy beta particle and transforms to stable, nonradioactive helium. Tritium occurs naturally in the environment in very low concentrations in the form of tritiated water, which easily disburses in the atmosphere, bodies of water, soil, and rock. Tritium is also a component in the triggering mechanism in thermonuclear (fusion) weapons. It is used in various self-luminescent devices, such as exit signs in buildings, aircraft dials, gauges, luminous paints, and wristwatches. Tritium primarily enters the body when people swallow tritiated water. People may also inhale tritium as a gas in the air and absorb it through their skin. Once tritium enters the body, it disperses quickly and is uniformly distributed throughout the body. Tritium is excreted through the urine within a month or so after ingestion. Organically bound tritium (tritium that is incorporated in organic compounds) can remain in the body for a longer period. Since tritium is almost always found as water, it goes directly into soft tissues and organs. It produces a low-level exposure and may result in toxic effects to the kidney. However, as with all ionizing radiation, exposure to tritium increases the risk of developing cancer.

Trust Territory of the Pacific Islands (TTPI): A United Nations trust territory established after World War II and administered by the United States beginning on July 18, 1947. The TTPI was a "strategic" trust territory, which meant that the United States did not have to disclose details about its military operations within the area.

units of radiation: The basic unit of radiation exposure is the roentgen, named after Wilhelm Roentgen (discoverer of x-rays). It is a measure of ionization in air, technically equal to one ESU (electrostatic unit) per cubic centimeter, due to radiation. A rep (roentgen equivalent physical) is an archaic measure of skin exposure to a dose of beta radiation having an effect equivalent to 1 roentgen of x-rays. The basic unit of radiation absorbed by the body is the rad, technically equal to 100 ergs (energy unit) per gram of exposed tissue. One roentgen corresponds to roughly 0.95 rad. The rem (roentgen equivalent in man) is a unit of effective dose, a dose corrected for the varying biological effectiveness of various types of ionizing radiation. The currently accepted unit of radiation is the gray (Gy), the International System unit of absorbed dose, equal to the energy imparted by ionizing radiation to a mass of matter corresponding to 1 joule per kilogram.

units of radioactivity: The becquerel (Bq), named after the physicist Henri Becquerel (the discoverer of radioactivity), is a measure of radioactivity equal to one atomic disintegration per second. The curie (Ci), whose name honors the French scientists Marie and Pierre Curie (the discoverers of radium), is a standard based on the radioactivity of 1 gram of radium. It is equal to 3.7 x 1010 becquerels.

walap: A large, oceangoing outrigger. The Marshallese, and particularly the people from Enewetak and Ujelang, were famous in the Pacific region for the construction and use of these vessels.

weto: A land parcel running the width of an island, including access to the ocean and the lagoon.

x-rays: Invisible, highly penetrating electromagnetic radiation of a much shorter wavelength than visible light, discovered in 1895 by Wilhelm C. Roentgen. Most applications of x-rays are based on their ability to pass through matter. They are dangerous in that they can destroy living tissue, causing severe skin burns on human flesh exposed for too long. This property is applied in x-ray therapy to destroy diseased cells. See also *ionizing radiation*.

Glossary Sources

Advisory Commission on Human Radiation Experimentation (ACHRE) Glossary
http://lsda.jsc.nasa.gov/common/gloss_se.cfm

Toxicological Profile Information Sheets, Glossary
Agency for Toxic Substances and Disease Registry, Department of Health and Human Services http://www.atsdr.cdc.gov/toxpro2.html

Health Effects of Project SHAD Chemical Agent: Phosphorus-32
Prepared for the National Academic by the Center for Research Information, Inc.
http://www.iom.edu/Object.File/Master/43/450/PHOSPHORUS%20-32.pdf

Marshall Islands: Glossary of Terms
Marshall Islands Dose Assessment and Radioecology Program https://eed.llnl.gov/mi/glossary.php

Mark Merlin, Alfred Capelle, Thomas Keene, James Juvik, and James Maragos, *Plants and Environments of the Marshall Islands* (Honolulu: East-West Center, 1997) http://www.hawaii.edu/cpis/MI/plants/bob.html

Radionuclides in the Environment—Radiation Glossary
United States Environmental Protection Agency http://www.epa.gov/rpdweb00/radionuclides/index.html

Medline Plus
U.S. National Library of Medicine and the National Institutes of Health http://www.nlm.nih.gov/medlineplus/ency/article/000909.htm

Index

*Numbers in **bold** refer to numbered plates in photo essay following page 56.*